Lecture Notes on Data Engineering and Communications Technologies

Volume 32

Series Editor

Fatos Xhafa, Technical University of Catalonia, Barcelona, Spain

The aim of the book series is to present cutting edge engineering approaches to data technologies and communications. It will publish latest advances on the engineering task of building and deploying distributed, scalable and reliable data infrastructures and communication systems.

The series will have a prominent applied focus on data technologies and communications with aim to promote the bridging from fundamental research on data science and networking to data engineering and communications that lead to industry products, business knowledge and standardisation.

** Indexing: The books of this series are submitted to ISI Proceedings, MetaPress, Springerlink and DBLP **

More information about this series at http://www.springer.com/series/15362

Jude Hemanth · Madhulika Bhatia ·
Oana Geman
Editors

Data Visualization
and Knowledge Engineering

Spotting Data Points with Artificial
Intelligence

 Springer

Editors
Jude Hemanth
Department of Electronics and
Communication Engineering (ECE)
Karunya University
Coimbatore, Tamil Nadu, India

Madhulika Bhatia
Amity University
Noida, Uttar Pradesh, India

Oana Geman
Department of Electrical Engineering
and Computer Science
Ştefan cel Mare University of Suceava
Suceava, Romania

ISSN 2367-4512 ISSN 2367-4520 (electronic)
Lecture Notes on Data Engineering and Communications Technologies
ISBN 978-3-030-25796-5 ISBN 978-3-030-25797-2 (eBook)
https://doi.org/10.1007/978-3-030-25797-2

This Springer imprint is published by the registered company Springer Nature Switzerland AG
The registered company address is: Gewerbestrasse 11, 6330 Cham, Switzerland

Contents

Cross Projects Defect Prediction Modeling

Lipika Goel and Sonam Gupta

Abstract Software defect prediction has been much studied in the field of research in Software Engineering. Within project Software defect prediction works well as there is sufficient amount of data available to train any model. But rarely local training data of the projects is available for predictions. There are many public defect data repositories available from various organizations. This availability leads to the motivation for Cross projects defect prediction. This chapter cites on defect prediction using cross projects defect data. We proposed two experiments with cross projects homogeneous metric set data and within projects data on open source software projects with class level information. The machine learning models including the ensemble approaches are used for prediction. The class imbalance problem is addressed using oversampling techniques. An empirical analysis is carried out to validate the performance of the models. The results indicate that cross projects defect prediction with homogeneous metric sets are comparable to within project defect prediction with statistical significance.

Keywords Defect prediction · Cross projects · Within-project · Machine learning · Class imbalance

1 Introduction

Defect prediction is an important topic of research in the field of software engineering. It has been under heavy investigation in recent years as a means to focus quality assurance. The results of defect prediction provide a comprehensive list of defect-prone source code artefacts. This list can be used by quality assurance team to allocate limited resources in an effective manner for validating software products so that more

L. Goel · S. Gupta (✉)
Ajay Kumar Garg Engineering College, Ghaziabad, India
e-mail: guptasonam@akgec.ac.in

L. Goel
e-mail: goellipika@akgec.ac.in

© Springer Nature Switzerland AG 2020
J. Hemanth et al. (eds.), *Data Visualization and Knowledge Engineering*,
Lecture Notes on Data Engineering and Communications Technologies 32,
https://doi.org/10.1007/978-3-030-25797-2_1

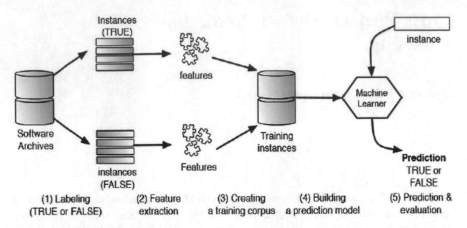

Fig. 1 General defect prediction process [1]

focus and effort could be provided on the defect-prone source code. There are various defect prediction techniques and most of the defect prediction models are based on machine learning. Within project defect prediction was a traditional defect prediction approach. Figure 1 explains the within project defect prediction (WPDP) approach.

In the previous years, Cross-Project Defect Prediction (CPDF) has grown from being an ignored point into a quite large sub-topic of software defect prediction [2, 3]. As the name itself suggests, by Cross-Project Defect Prediction we mean predicting the defects using the training data from some other projects. When the source and the target projects are different but they have the same set of metrics then it is called homogeneous defect predict [4, 5]. When the source and the target projects are different and they have the different set of metrics then it is called heterogeneous defect predict [6, 7]. In a study [8], it was found that around 50 publications specifically addressed the cross-project aspects between 2012 and 2015. However, from the survey it was also concluded that it is not easy to compare which of the many proposed approaches performs the best.

In the state of art four different categories of cross projects have been discussed [9, 10]. Figure 2 gives diagrammatic representation of it.

- Strict Cross Project Defect Prediction: The training data does not contain any information of the target project.
- Mixed Cross Project Defect Prediction: The training data will contain data of the older versions of the target project.
- Mixed Cross Project Defect Prediction with target class data: The training data will contain some data from target class data.
- Pairwise Cross Project Defect Prediction: In this the project is trained on one project and tested on the other target project.

In this chapter we have performed a systematic analysis of the performance of the classifiers for CPDP. For handling the class imbalance problem SMOTE tech-

Fig. 2 Different categories
of CPDP (*Source*
Herbold [8])

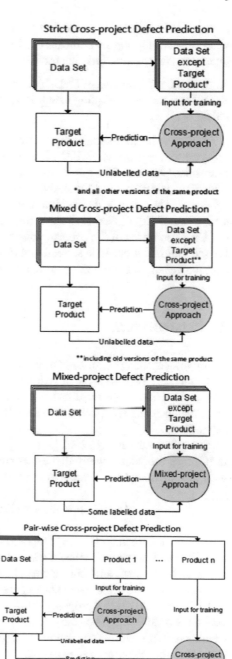

nique of oversampling is performed and is evaluated. The results of CPDP with and without oversampling are compared with within project defect prediction. The major questions addressed in this chapter are:

RQ1. Is the performance of CPDP comparable to within project defect prediction. RQ2. Does the oversampling using SMOTE improves the performance of the classifier?

To answer he above stated research questions we have conducted the experiment on open source projects. SMOTE is used to oversample the data to handle the class imbalance problem. The results were compared with within project defect prediction. The classifiers used for study are logistic regression, decision tree, SVM and XGBoost. The paper is divided into the following sections, Sect. 2 discusses about state of art of CPDP. Section 3 states the datasets, the models used of for defect prediction, description of the oversampling technique and performance measures used for the evaluation of models. Section 4 describes the methodology used. Section 5 summarizes the results. Section 6 concludes the paper.

2   Related Work

Khoshgoftaar et al. [11] in 2008 proposed to use a classifier that can be used to train each and every product separately and then use the majority vote that is obtained by all the classifiers to perform further classification. These authors also investigated other methods and approaches; however, this majority vote approach was best in the suggestions.

Even Watanabe et al. in 2008 [12], proposed to standardize the target data in order to improve the homogeneity that exists between the training and target data. They used this mathematical formula to standardize target data given as:

$$mi = mi\,(s) * .mean\,(mi\,(S)) * /mean\,(mi\,(S)) * .$$

Turhan et al. in 2009 [13, 14] proposed to first transform the metric data with the logarithm and then applied the relevancy filter to the training data that was available based on k-Nearest Neighbour algorithm. Using the relevancy filter, the k nearest instance to each target is selected and generally the value of k is taken as 10.

Liu et al. in 2010 [15], proposed the S-expression trees that were trained using the genetic program which used the validation and voting strategy. It used the parts of the training data for internal validation and majority votes of multiple genetic programs were used for the purpose of classification.

Menzies et al. in 2011 [16] and 2013 [17], created a local model which involved the clustering of the training data with WHERE algorithm and then whatever results that were obtained were classified using the WHICH rule learning algorithm.

Herbold, in 2013 [18] proposed to use the KNN algorithm for the suitable filtering of products. This KNN algorithm uses the distance between the distributional characteristics of products to find out the neighbours. They suggested to put k = floor value of (#product/2) of the products for training. They even used the relevancy filters using the EM clustering algorithm, but thus approach was surpassed by the KNN algorithm.

Ryu et al. in 2014 [19] used an approach based on resampling. Here, firstly 10% of the training data is selected randomly in order to evaluate the boosting result. Then for each boosting iteration, based on the similarity the data is resampled. Here similarity was determined using the data weighting approach that was proposed by Ma et al. [20].

In 2015, Peters [21] proposed LACE2 as an extension of CLIFF and MORPH for privatization. The authors introduced the concept of shared cache, in which the data owners could add their data one after the another. Any new data was added if it was not already represented in the data. So, the authors proposed a variant to the leader—follower algorithm.

Y. Zhang et al. also in 2015 [22] presented the use of the ensemble classifiers Average and Maximum voting, which used Alternating Decision Trees, Bayesian networks, Logistics regression, RBF networks etc. They even proposed to use.

Bagging [23] and Boosting [24] for Naïve Bayes and Logistic Regression classifier. Xin Xia et al. in 2016 [25] proposed Hybrid Model Reconstruction Approach also known as HYDRA for CPDP modeling. In this model the genetic algorithm and the ensemble learning approach is used for building the classifier. The HYDRA outperformed than the existing models in the state of art.

Zhang et al. [26] discussed about heterogeneous cross project defect prediction (HCPDP). The results were comparable to supervised learning. The models made good prediction on cross projects.

To benchmark the different approaches of cross project defect prediction, Steffen Herbold et al. in 2017 [8] did a comparative study. He stated that CPDP has still not reached to a level where it can be used for the application in practice.

3 Prerequisite Knowledge

3.1 Dataset and Metrics

The datasets used for the study are taken from publicly available PROMISE repository for NASA related projects and http://openscience.com. The datasets considered are imbalanced in nature. There exists an imbalance in the number of defect prone and non defect prone classes. Table 1 gives the description of the datasets.

For cross project prediction, we have considered those projects that have same set of metrics thereby performing homogeneous cross project defect prediction. We have considered the CK metrics for software defect prediction [27–30]. CK metrics gives software metrics for object-oriented (OO) software implementation. The description of the CK metrics is given in Table 2. The list of CK metrics are:

- WMC (weighted methods per class)
- DIT (depth of inheritance tree)
- NOC (number of children)
- CBO (coupling between objects)
- RFC (response for class)
- LCOM (lack of cohesion of methods).

Table 1 Description of datasets

Project	Total no of classes	No of Non defect prone classes	No of defect prone classes
Synapse 1.0	157	144	13
Synapse1.2	223	161	62
Xalan 1.2	723	613	110
Xereces 1.2	440	369	71

Table 2 Description of the dataset

CK metrics	Description
WMC	The combined complexity of the different methods of the class is called WMC. It gives the prediction of the efforts and the time required to develop and maintain a class. Higher the WMC higher is the complexity of the class
DIT	The maximum length of a tree from the root node to the specific node. Higher the value of DIT higher is the reusability
NOC	Number of subclass of a class is called NOC. Greater the value greater is the reusability and greater is the testing efforts required
CBO	The number of classes coupled to a class is called CBO. Higher the value higher is the complexity and higher is the testing efforts
RFC	The total number of methods of a class is called RFC. Higher the value higher is the complexity of the class
LCOM	The degree of dissimilarity of methods in a class is called LCOM. Higher the value higher is the complexity and higher are the efforts required for testing the class

3.2 Description of Classifiers

Logistic Regression

Logistic Regression is the regression analysis which is conducted when the dependent variable is dichotomous (binary). A binary logistic has a dependent variable with possibly two values like pass/fail, win/loss, white/black or true/false etc. It describes the data and even explains the relationship between the single dependent binary variable and one or more nominal independent variables. Some assumptions that logistic regression uses include that the variable must be dichotomous in nature (i.e. present or absent) [31, 32]. The predictors should not have high degree of correlation. This can be assessed by creating a correlation matrix amongst the predictors. It is used to determine whether some email is spam or not, or if some tumour is malignant or not. Sometimes this logistics regression is difficult to conduct however the Intellectus Statistics tools easily allow us to perform the regression analysis.

SVM

SVM or Support vector machine is discriminative classifier that classifies the data using a separating hyperplane. It is a supervised learning model that is used to analyse the data for classification and regression analysis [33]. Support vectors are the data points that are closer to the hyperplane and influence the position and orientation of the hyperplane. Here we plot each data as a point in the n-dimensional space (where n corresponds to the n features that are to be considered). Then classification that is used to separate the two classes by the hyperplane. The main points that are to be remembered are that we have to select a hyperplane that segregates the two classes in a more better and efficient manner by maximising the distance between the nearest data point and the hyperplane. For example, a hyperplane is selected to separate the points in the input variable space into two classes, either class 0 or class 1. In two-dimensions we can visualize this as a line. Let us assume that all of our input points can be completely separated by this line. For example:

$$B0 + (B1 * X1) + (B2 * X2) = 0$$

where the coefficients (B1 and B2) that determine the slope of the line and the intercept (B0) are found by the learning algorithm, and X1 and X2 are the two input variables.

XGBoost

It is a open source library that is used for providing the gradient boosting framework for C++, Python, R, Julia and Java [34]. It can work on windows, Linux as well as MacOS. The aim of XGBoost I to provide scalable, portable and distributed gradient boosting libraries for the above mentioned frameworks. One of the most important feature includes the scalability, which drives fast learning through parallel and distributed computing. This even and offers efficient memory usage. XGBoost is an implementation of gradient boosted decision trees. These are also designed for high speed and performance, and hence it is referred as a scalable tree boosting system. The algorithm was developed to efficiently reduce the computing time and allocate an optimal usage of memory resources. Important features for the implementation include handling of the missing values, Block Structure to support parallelization in tree construction and the ability to fit and boost on new data added to a trained model (Continued Training).

Decision trees

Decision tree is one of the most powerful and popular tool for classification and prediction. A Decision tree is like tree structure, where each internal node denotes a test on an attribute, each branch represents an outcome of the test, and each leaf node (terminal node) holds a class label [35]. A decision tree consists of three types of nodes:

- Decision nodes—typically represented by squares and represents the condition or the decision that is to be taken.
- Chance nodes—typically represented by circles.
- End nodes—typically represented by triangles.

Lets say, we are predicting the price of houses. Now the decision tree will start splitting after we have considered all the features in the training data. The mean of responses of the training data inputs of particular group is considered as prediction for that group. The above function is applied to all data points and cost is calculated for all candidate splits. Again the split which has the lowest cost is chosen. Another cost function involves reduction of standard deviation.

3.3   Class Imbalance Problem

SMOTE Algorithm

In case of pattern recognition and data mining, Imbalanced data classification often arises. Using a machine learning algorithm out of the box is problematic when one class in the training set dominates the other [36]. Synthetic Minority Over-sampling Technique (SMOTE) solves this problem.

There are 4 ways of addressing this class imbalance problems by:

- Synthesising the of new minority class instances
- Over-sampling of the minority class
- Under-sampling of the majority class

- Tweak the cost function to make misclassification of minority instances more important than misclassification of majority instances.

SMOTE synthesizes the new minority instances between the minority instances that already exist. The SMOTE() of smote family takes two parameters: K and dup_size. This SMOTE() function thinks from the perspective of existing minority instances and then synthesizes the new instances which are at some distance from them, If the value of K is taken as 1 then only the closest neighbor of the same class is considered. If the value of K is 2, then the closest and the second closest neighbor of the same class is to be considered. The dup_size parameter answers the question that how many times SMOTE() function should loop through the existing instances that were in minority.

4 Performance Measures

Performance Measures are used to evaluate the performance of the models. The confusion matrix is generated [37, 38]. The standard confusion matrix is given in Fig. 3. The following are the parameters of the confusion matrix:

- True positive: fault prone classes predicted as fault prone.
- False positive: non fault prone classes predicted as fault prone.
- True negative: non fault prone classes predicted as non fault prone.
- False negative: fault prone instances predicted as non fault prone.

Using these parameters of confusion matrix the following performance measures are evaluated:

- Precision = (TP)/(TP + FP)
- Recall = (TP)/(TP + FN)
- Accuracy = (TP + TN)/(TP + FP + TN + FN)
- F1 score = (2 * Precision * Recall)/(Precision + Recall)
- AUC-ROC value: It quantifies the overall effectiveness of the various algorithms.

The description of the performance measures are in Table 3.

	Fault-Prone	Non-Fault Prone
Non Faulty	FP	TN
Faulty	TP	FN

Fig. 3 Standard confusion matrix

Table 3 Description of the performance measures

Performance measures	Description
Precision	It is the degree to which repeated experiments under the same conditions show the identical results
Recall	It is the measure of correctly predicted fault prone instances amon all the actual fault prone instances
F-Score	It is the harmonic mean of precision and recall
Accuracy	The degree of correctly classified defect and non defect prone instances
AUC-ROC	It is the trade off between false alarm rate and true positive rate. It measures the effectiveness of the model

5 Methodology

The main objective of the experiment performed is two fold. Firstly we wanted to identify that if the performance of the cross project defect prediction process is comparable to within project defect prediction process. Secondly, since the datasets were imbalanced therefore we wanted to analyze whether application of data sampling using SMOTE technique improves the predictive performance of the model.

To accomplish theses two objectives we have conducted two experiments.

Experiment 1

Different combinations of the datasets except the target project are taken into consideration for training the model. The preprocessing involves removing the redundant data among all datasets, label encoding of the categorical data and extraction of the CK Metrics from the datasets.

All the classifiers are used for each of the combination. Using the confusion matrix the performance measures are evaluated and the performance of the model is analyzed. In this scenario defect prediction is done on the imbalanced data. No data sampling is technique is applied for balancing the training data.

The results of the defect prediction are compared with within project defect prediction. The same target project is also used for training the model in a 70:30 training testing ratio. The same classifiers are used and the results are compared under the same performance measures.

Experiment 2

Same as in Experiment 1 different combinations of the datasets except the target project are taken into consideration for training the model. The preprocessing involves removing the redundant data among all datasets, label encoding of the categorical data and extraction of the CK Metrics from the datasets. In this scenario data sampling is performed using SMOTE technique to handle the class imbalance problem. On application of the SMOTE technique their will be balance in the number of defect prone and non defect prone classes. Then the training dataset is provided to the model for prediction.

All the classifiers are used for each of the combination. Using the confusion matrix the performance measures are evaluated and the performance of the model is analyzed.

Same as in Experiment 1 the results of the defect prediction are compared with within project defect prediction. For within project defect prediction also the data sampling technique is applied to handle the class imbalance problem. Then the balanced training data is passed to the model for the prediction. The same target project is also used for training the model in a 70:30 training testing ratio. The same classifiers are used and the results are compared under the same performance measures. Figures 4 and 5 explains the training and testing data set selection of CPDP and WPDP respectively.

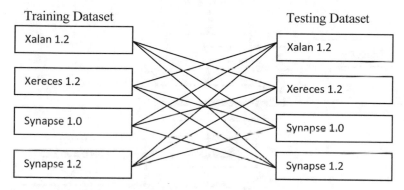

Fig. 4 Training and testing dataset selection in cross project defect prediction

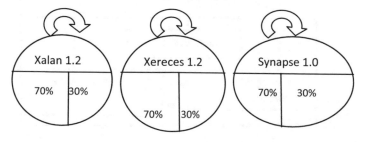

Fig. 5 Training and testing dataset selection in within project defect prediction

Figures 6 and 7 gives the description of the defect prediction process used with and without sampling.

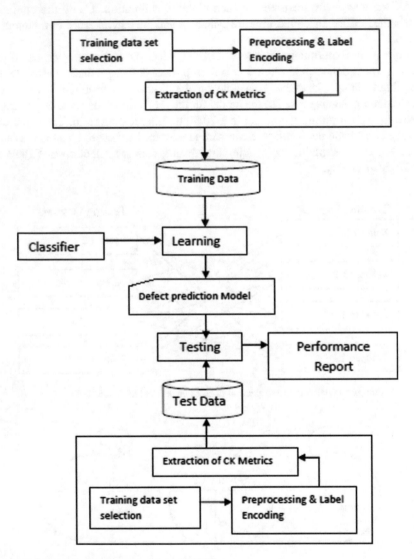

Fig. 6 Defect prediction process without SMOTE

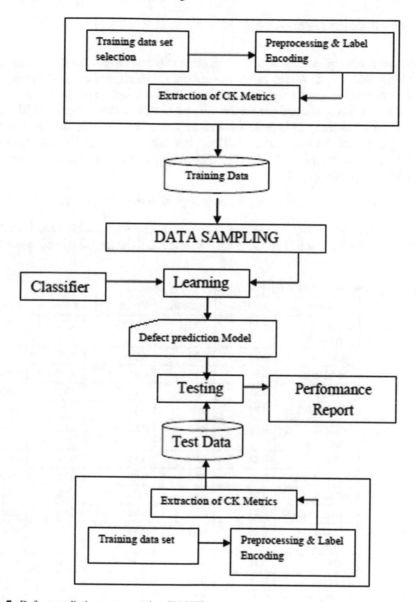

Fig. 7 Defect prediction process using SMOTE

6 Results and Discussion

In this section we present results of the experiments conducted. We summarize our results in Tables 4, 5, 6 and 7. We use precision, recall, accuracy, f1-score and ROC as the performance evaluation measures. Tables 2 and 3 are the results of the Experiment 1. These two tables tabulates the results without data sampling (SMOTE) applied for cross and within project defect prediction respectively. Tables 4 and 5 are the results of the Experiment 2. These two tables tabulates the results when data sampling technique ie. SMOTE is applied for cross and within project defect prediction respectively.

RQ1. Whether CPDP is comparable to within project defect prediction?

From Table 4 we can state that the accuracy of Cross prediction defect prediction is 0.77, 0.83 and 0.81 for Xalan 2.4., Xerces1.2, Synapse 1.0 and Synapse 1.1

Table 4 Results without SMOTE for cross project defect prediction

Testing	Classifier	Accuracy	Precision	Recall (PD)	F1-score	ROC
Xalan 2.4	Logistic regression	0.79	0.75	0.81	0.77	0.510
	SVM	0.74	0.74	0.78	0.75	0.531
	Decision tree	0.73	0.76	0.81	0.77	0.594
	XG boost	0.81	0.75	0.82	0.77	0.590
	Average values	0.77	0.75	0.81	0.77	0.556
Xerces1.2	Logistic regression	0.83	0.86	0.82	0.83	0.641
	SVM	0.83	0.85	0.82	0.82	0.639
	Decision tree	0.81	0.85	0.86	0.85	0.678
	XG boost	0.85	0.87	0.86	0.86	0.632
	Average values	0.83	0.86	0.84	0.84	0.648
Synapse 1.0	Logistic regression	0.79	0.79	0.81	0.8	0.621
	SVM	0.78	0.74	0.78	0.76	0.651
	Decision tree	0.85	0.79	0.82	0.8	0.601
	XG boost	0.8	0.79	0.81	0.8	0.612
	Average values	0.81	0.78	0.81	0.78	0.625
Synapse 1.1	Logistic regression	0.85	0.87	0.88	0.87	0.621
	SVM	0.83	0.85	0.85	0.85	0.632
	Decision tree	0.85	0.88	0.86	0.87	0.589
	XG boost	0.86	0.87	0.89	0.88	0.635
	Average values	0.85	0.87	0.87	0.87	0.614

Table 5 Results without SMOTE for within project defect prediction

Testing	Classifier	Accuracy	Precision	Recall (PD)	F1-score	ROC
Xalan 2.4	Logistic regression	0.79	0.75	0.83	0.77	0.512
	SVM	0.75	0.75	0.78	0.76	0.523
	Decision tree	0.75	0.77	0.82	0.80	0.597
	XG boost	0.82	0.78	0.83	0.79	0.612
	Average values	0.78	0.76	0.82	0.78	0.561
Xerces1.2	Logistic regression	0.85	0.86	0.83	0.84	0.645
	SVM	0.86	0.88	0.85	0.86	0.656
	Decision tree	0.83	0.87	0.88	0.87	0.687
	XG boost	0.87	0.89	0.87	0.88	0.686
	Average values	0.85	0.88	0.86	0.86	0.669
Synapse 1.0	Logistic regression	0.8	0.81	0.83	0.82	0.632
	SVM	0.8	0.78	0.81	0.79	0.651
	Decision tree	0.86	0.81	0.83	0.82	0.613
	XG boost	0.81	0.82	0.84	0.83	0.615
	Average values	0.82	0.81	0.83	0.82	0.628
Synapse 1.1	Logistic regression	0.88	0.88	0.88	0.88	0.685
	SVM	0.84	0.86	0.88	0.86	0.656
	Decision tree	0.60	0.88	0.86	0.87	0.612
	XG boost	0.87	0.87	0.91	0.89	0.721
	Average values	0.80	0.87	0.88	0.88	0.669

respectively. These values are comparable to the accuracy of within project defect prediction which is 0.78, 0.85, 0.82 and 0.80 for Xalan 2.4., Xerces1.2, Synapse 1.0 and Synapse 1.1 respectively.

The value of precision in CPDP is 0.75, 0.86, 0.78 and 0.87 Xalan 2.4., Xerces1.2, Synapse 1.0 and Synapse 1.1 respectively whereas the value for precision in within project is 0.76, 0.88, 0.81 0.87 for the above mentioned projects.

The F1 score is also comparable to within project values. The F1-score value for CPDP is 0.77, 0.84, 0.78 and 0.87 for Xalan 2.4., Xerces1.2, Synapse 1.0 and Synapse 1.1 respectively whereas for WPDP it is 0.78, 0.86, 0.82 and 0.88 for XGBoost for the above mentioned projects.

The highest value of AUC-ROC in CPDP is 0.614 whereas in WDP it is 0.669. On empirical evaluation of these results we can infer that the performance of CPDP is comparable to Within Project defect prediction.

RQ2. Does the oversampling using SMOTE improves the performance of the classi-fier?

From Tables 6 and 7 we can infer that on using SMOTE the accuracy of Cross prediction defect prediction is 0.81, 0.84, 0.81 and 0.88 for Xalan 2.4., Xerces1.2, Synapse 1.0 and Synapse 1.1 respectively. These values are higher by approximately 2.45% to the accuracy of projects without data sampling (SMOTE).

The value of precision in CPDP is 0.77, 0.86, 0.78 and 0.90 Xalan 2.4., Xerces1.2, Synapse 1.0 and Synapse 1.1 respectively. These values are also higher by approxi-mately 1.53% to the precision of projects without data sampling (SMOTE).

The Recall value for CPDP is 0.83, 0.85, 0.81 and 0.89 for Xalan 2.4., Xerces1.2, Synapse 1.0 and Synapse 1.1. This is also higher by approximately 1.5% to the recall of projects without data sampling (SMOTE).

The F1 score being the harmonic mean of precision and recall also showed similar performance. The value of F-score in CPDP is 0.77, 0.84, 0.78 and 0.88 for Xalan 2.4.,

Table 6 Results using SMOTE for cross project defect prediction

Testing	Classifier	Accuracy	Precision	Recall (PD)	F1-score	ROC
Xalan 2.4	Logistic regression	0.79	0.78	0.83	0.77	0.511
	SVM	0.81	0.77	0.82	0.75	0.551
	Decision tree	0.77	0.78	0.83	0.78	0.594
	XG boost	0.86	0.76	0.83	0.78	0.591
	Average values	0.81	0.77	0.83	0.77	0.562
Xerces1.2	Logistic regression	0.83	0.86	0.82	0.86	0.641
	SVM	0.84	0.85	0.84	0.81	0.639
	Decision tree	0.84	0.87	0.86	0.86	0.696
	XG boost	0.86	0.87	0.86	0.81	0.732
	Average values	0.84	0.86	0.85	0.84	0.677
Synapse 1.0	Logistic regression	0.81	0.79	0.81	0.78	0.635
	SVM	0.82	0.74	0.78	0.75	0.590
	Decision tree	0.81	0.79	0.82	0.79	0.638
	XG boost	0.808	0.79	0.81	0.78	0.636
	Average values	0.81	0.78	0.81	0.78	0.625
Synapse 1.1	Logistic regression	0.89	0.90	0.90	0.90	0.60
	SVM	0.87	0.90	0.87	0.85	0.64
	Decision tree	0.87	0.89	0.89	0.88	0.56
	XG boost	0.89	0.91	0.90	0.90	0.65
	Average values	0.88	0.90	0.89	0.88	0.61

Table 7 Results using SMOTE for within project defect prediction

Testing	Classifier	Accuracy	Precision	Recall (PD)	F1-score	ROC
Xalan 2.4	Logistic regression	0.79	0.79	0.85	0.82	0.623
	SVM	0.78	0.77	0.84	0.81	0.621
	Decision tree	0.8	0.76	0.83	0.8	0.695
	XG boost	0.89	0.84	0.86	0.85	0.601
	Average values	0.82	0.79	0.85	0.82	0.635
Xerces1.2	Logistic regression	0.85	0.86	0.88	0.86	0.641
	SVM	0.84	0.89	0.86	0.87	0.687
	Decision tree	0.87	0.92	0.86	0.88	0.654
	XG boost	0.86	0.87	0.87	0.87	0.754
	Average values	0.86	0.89	0.87	0.87	0.684
Synapse 1.0	Logistic regression	0.81	0.79	0.81	0.78	0.635
	SVM	0.87	0.78	0.8	0.79	0.697
	Decision tree	0.82	0.79	0.82	0.81	0.654
	XG boost	0.84	0.79	0.85	0.82	0.632
	Average values	0.84	0.79	0.82	0.80	0.655
Synapse 1.1	Logistic regression	0.89	0.90	0.90	0.90	0.61
	SVM	0.88	0.92	0.87	0.88	0.64
	Decision tree	0.87	0.96	0.87	0.91	0.56
	XG boost	0.92	0.91	0.92	0.91	0.65
	Average values	0.89	0.92	0.89	0.90	0.62

Xerces1.2, Synapse 1.0 and Synapse 1.1 respectively. These values are also higher by approximately to the precision of projects without data sampling (SMOTE).

The AUC-ROC value also showed an improvement after using SMOTE. The improvement in the values of Accuracy, Precision, Recall, F-Score and AUC-ROC on using data sampling for CPDP than predictions with no data sampling infer that data sampling in CPDP is effective for improving the predictive performance of the model.

CPDP is comparable to within project defect prediction with statistical significance.

SMOTE technique of oversampling improves the predictive performance
of the model.

7   Threats to Validity

The treats to validity is categorized under the criteria of threats to internal validity, threats to external validity, threats to reliability validity and threats to conclusion validity.

Threats to internal validity describes the mistakes or the errors in the experiments [39, 40]. There might be some errors in the experiment that went unnoticed. The use of classifier in modelling can effect the internal validity. There are many classification algorithms available. The quality issues in the datasets can be a concern or threat to internal validity.

The generalization of the results might result in threats to external validity [41]. The datasets considered for the experiment are from open source software and from single source. On consideration of more datasets from different sources we can reduce the threat of external validity.

The replication of the work leads to the threats to reliability validity [42, 43]. Since the datasets used are publicly available and the implementation is available therefore they may be a threat to reliability.

The threat to conclusion validity is that compared the results with WPDP to prove the feasibility of classification for CPDP. The performance measures used are recall, precision, F-score and AUC values. Some statistical test or cost effectiveness measures can also be considered for reducing the threat of conclusion validity.

8   Conclusion

The cross project defect prediction (CPDP) is the most upcoming and trending research area in the field of software engineering. It improves the software quality and reliability. The testing efforts, time and cost of the project is reduced to an extent using defect prediction. A more focussed testing increases the probability of an error free software. CPDP enables the prediction of defects even when little or no historical information of defects is available for the software. Imbalance in the defect prone and the non defect prone classes degrades the performance of the model.

In this paper we have tried to handle the class imbalance problem using SMOTE technique of oversampling. We conducted experiments on cross projects and within project data using four open source projects. The results proved that the performance of cross projects defect prediction is comparable to within project defect prediction with statistical significance. The results also stated that data sampling using SMOTE technique in CPDP is effective and it improved the predictive performance of the model. From the experiment it was also evident that CK based object oriented metrics are effective set of features for cross project defect prediction.

In future work, we would like to extend our above stated conclusions on more data from different projects. Further work can be carried out for solving the of class imbalance problem for the projects with different set of features. A novel approach of data sampling can be proposed for more real time applications.

References

1. Kim S, Zhang H, Wu R, Gong L (2011) Dealing with noise in defect prediction. In: 33rd international conference on software engineering (ICSE). Waikiki, Honolulu, HI, USA: ACM, 978-1-4503-0445
2. Uchigaki S, Uchida S, Toda K, Monden A (2012) An ensemble approach of simple regression models to cross-project fault prediction. In: 2012 13th ACIS international conference on software engineering, artificial intelligence, networking and parallel distributed computing (SNPD), Aug 2012, pp 476–481
3. He Z, Peters F, Menzies T, Yang Y (2013) Learning from opensource projects: an empirical study on defect prediction. In: Proceedings of the 7th international symposium on empirical software engineering and measurement (ESEM)
4. Canfora G, Lucia AD, Penta MD, Oliveto R, Panichella A, Panichella S (2013) Multi-objective cross-project defect prediction. In: Proceedings of the 6th IEEE international conference on software testing, verification and validation (ICST)
5. Peters F, Menzies T, Gong L, Zhang H (2013) Balancing privacy and utility in cross-company defect prediction. IEEE Trans Softw Eng 39(8):1054–1068
6. Nam J, Kim S (2015) Heterogeneous defect prediction. In: Proceedings of the 2015 10th joint meeting on foundations of software engineering, series ESEC/FSE 2015. ACM, New York, NY, USA, pp 508–519 [Online]. Available: http://doi.acm.org/10.1145/2786805.2786814
7. Jing X, Wu F, Dong X, Qi F, Xu B (2015) Heterogeneous crosscompany defect prediction by unified metric representation and CCA-based transfer learning. In: Proceedings of the 2015 10th joint meeting on foundations of software engineering, series ESEC/FSE 2015. ACM, New York, NY, USA, pp 496–507 [Online]. Available: http://doi.acm.org/10.1145/2786805. 2786813
8. Herbold S (2017) A systematic mapping study on cross-project defect prediction. CoRR, vol. abs/1705.06429, 2017 [Online]. Available: https://arxiv.org/abs/1705.06429
9. Jureczko M, Madeyski L (2010) Towards identifying software project clusters with regard to defect prediction. In: Proceedings of the 6th international conference on predictive models in software engineering (PROMISE). ACM
10. Jureczko M, Madeyski L (2015) Cross–project defect prediction with respect to code ownership model: an empirical study. e-Inform Softw Eng J 9:21–35
11. Khoshgoftaar TM, Rebours P, Seliya N (2008) Software quality analysis by combining multiple projects and learners. Softw Qual J 17(1):25–49 [Online]. Available: http://dx.doi.org/10.1007/s11219-008-9058-3
12. Watanabe S, Kaiya H, Kaijiri K (2008) Adapting a fault prediction model to allow inter language reuse. In: Proceedings of the 4th international workshop on predictor models in software engineering (PROMISE). ACM
13. Turhan B, Tosun A, Bener A (2011) Empirical evaluation of mixed-project defect prediction models. In: 2011 37th EUROMICRO conference on software engineering and advanced applications (SEAA), pp 396–403

14. Turhan B, Menzies T, Bener A, Di Stefano J (2009) On the relative value of cross-company and within-company data for defect prediction. Empir Softw Eng 14:540–578
15. Liu Y, Khoshgoftaar T, Seliya N (2010) Evolutionary optimization of software quality modeling with multiple repositories. IEEE Trans Softw Eng 36(6):852–864
16. Menzies T, Butcher A, Marcus A, Zimmermann T, Cok D (2011) Local vs. global models for effort estimation and defect prediction. In: Proceedings of 26th IEEE/ACM international conference on automated software engineering (ASE). IEEE Computer Society
17. Menzies T, Butcher A, Cok D, Marcus A, Layman L, Shull F, Turhan B, Zimmermann T (2013) Local versus global lessons for defect prediction and effort estimation. IEEE Trans Softw Eng 39(6):822–834
18. Camargo Cruz AE, Ochimizu K (2009) Towards logistic regression models for predicting fault-prone code across software projects. In: Proceedings of the 3rd international symposium on empirical software engineering and measurement (ESEM). IEEE Computer Society
19. Ryu D, Choi O, Baik J (2014) Value-cognitive boosting with a support vector machine for cross-project defect prediction. Empir Softw Eng 21(1):43–71 [Online]. Available: http://dx.doi.org/10.1007/s10664-014-9346-4
20. Ma Y, Luo G, Zeng X, Chen A (2012) Transfer learning for cross-company software defect prediction. Inf Softw Technol 54(3):248–256
21. Peters F, Menzies T, Layman L (2015) LACE2: better privacy preserving data sharing for cross project defect prediction. In: 2015 IEEE/ACM 37th IEEE international conference on software engineering (ICSE), vol 1, pp 801–811
22. Zhang Y, Lo D, Xia X, Sun J (2015) An empirical study of classifier combination for cross-project defect prediction. In: 2015 IEEE 39th annual computer software and applications conference (COMPSAC), vol 2, pp 264–269
23. Breiman L (1996) Bagging predictors. Mach Learn 24(2):123–140
24. Freund Y, Schapire RE (1997) A decision-theoretic generalization of on-line learning and an application to boosting. J Comput Syst Sci 55(1):119–139 [Online]. Available: http://www.sciencedirect.com/science/article/pii/S002200009791504X
25. Xia X et al (2016) Hydra: massively compositional model for cross-project defect prediction. IEEE Trans Softw Eng 42(10):977–998
26. Zhang F et al (2016) Cross-project defect prediction using a connectivity- based unsupervised classifier. In: Proceedings of the 38th international conference on software engineering. ACM
27. Panichella A, Oliveto R, De Lucia A (2014) Cross-project defect prediction models: L'union fait la force. In: 2014 software evolution week—IEEE conference on software maintenance, reengineering and reverse engineering (CSMRWCRE), pp 164–173
28. Hall MA (1998) Correlation-based feature subset selection for machine learning. Ph.D. dissertation, University of Waikato, Hamilton, New Zealand
29. Briand LC, Melo WL, Wust J (2002) Assessing the applicability of fault-proneness models across object-oriented software projects. IEEE Trans Softw Eng 28(7):706–720
30. Singh P, Verma S, Vyas OP (2013) Article: cross company and within company fault prediction using object oriented metrics. Int J Comput Appl 74(8):5–11 (full text available)
31. Amasaki S, Kawata K, Yokogawa T (2015) Improving crossproject defect prediction methods with data simplification. In: 2015 41st Euromicro conference on software engineering and advanced applications (SEAA), Aug 2015, pp 96–103
32. He Z, Shu F, Yang Y, Li M, Wang Q (2012) An investigation on the feasibility of cross-project defect prediction. Autom Softw Eng 19:167–199
33. Raman B, Ioerger TR (2003) Enhancing learning using feature and example selection. Technical Report, Department of Computer Science, Texas A&M University
34. Ryu D, Jang J-I, Baik J (2015a) A transfer cost-sensitive boosting approach for cross-project defect prediction. Softw Qual J, pp 1–38 [Online]. Available: http://dx.doi.org/10.1007/s11219-015-9287-1
35. Zimmermann T, Nagappan N, Gall H, Giger E, Murphy B (2009) Cross-project defect prediction: a large scale experiment on data vs. domain vs. process. In: Proceedings of the 7th joint meeting of the european software engineering conference (ESEC) and the ACM SIGSOFT symposium on the foundations of software engineering (FSE). ACM, pp 91–100

36. Kawata K, Amasaki S, Yokogawa T (2015) Improving relevancy filter methods for cross-project defect prediction. In: 2015 3rd international conference on applied computing and information technology/2nd international conference on computational science and intelligence (ACIT-CSI), July 2015, pp 2–7
37. Nam J, Pan S, Kim S (2013) Transfer defect learning. In: 2013 35th international conference on software engineering (ICSE), pp 382–391
38. Ryu D, Jang J-I, Baik J (2015) A hybrid instance selection using nearest-neighbor for cross-project defect prediction. J Comput Sci Technol 30(5):969–980 [Online]. Available: http://dx.doi.org/10.1007/s11390-015-1575-5
39. Nam J, Kim S (2015) Clami: defect prediction on unlabeled datasets. In: 2015 30th IEEE/ACM international conference on automated software engineering (ASE), Nov 2015, pp 452–463
40. Peters F, Menzies T, Marcus A (2013) Better cross company defect prediction. In: Proceedings of the 10th working conference on mining software repositories, series MSR '13. IEEE Press, Piscataway, NJ, USA, pp 409–418 [Online]. Available: http://dl.acm.org/citation.cfm?id=2487085.2487161
41. Mizuno O, Hirata Y (2014) A cross-project evaluation of textbased fault-prone module prediction. In: 2014 6th international workshop on empirical software engineering in practice (IWESEP), Nov 2014, pp 43–48
42. Rahman F, Posnett D, Devanbu P (2012) Recalling the "imprecision" of cross-project defect prediction. In: Proceedings of the ACM SIGSOFT 20th international symposium on the foundations software engineering (FSE). ACM
43. Turhan B, Misirli AT, Bener A (2013) Empirical evaluation of the effects of mixed project data on learning defect predictors. Inf Softw Technol 55(6):1101–1118 [Online]. Available: http://doi.org/10.1016/j.infsof.2012.10.003

Recommendation Systems for Interactive Multimedia Entertainment

Shilpi Aggarwal, Dipanjan Goswami, Madhurima Hooda,
Amirta Chakravarty, Arpan Kar and Vasudha

Abstract In the era of internet has embarked on World Wide Web for searching or knowing any information about any topic which has resulted in humongous information load. Access to open source creates ever-growing content on web, where user will succumb to thousands of arrays of options available, when seeking for any product or services. As the entertainment industry is no way an exception, optimization of user's choice is of utmost importance. Recommendation system (RS) is the application that provides decision tool for tracking the user's previous choices about products or browsing web pages history, clicks and choices of similar users. The main purpose of RS is to support the user to select his desired product among the multiple, equally-competitive choices available. It can be employed for diverse set of item recommendation such as books, songs, movies, restaurants, gadgets, e-learning materials etc. This chapter attempts to explore the various techniques and reveals the human–machine cooperation by implementing recommendation systems, especially meant for entertainment industry. In summary, the essence of this treatise is the soft computing/knowledge engineering approaches (such as filtering techniques, machine learning algorithms) guided by internet of things to provide prediction that can be displayed by data visualization in entertainment landscape convincingly.

Keywords Recommendation engine · Entertainment · Machine learning · Internet of things · Predictive analytics · Knowledge engineering

S. Aggarwal (✉)
Manav Rachna International Institute of Research and Studies, Faridabad, India
e-mail: shilpi.16.g@amail.com

D. Goswami · Vasudha
IntuiComp TeraScience Pvt. Limited, New Delhi, India

M. Hooda
Amity University, Noida, India

A. Chakravarty
Sun Pharmaceutical Industries Limited, Gurugram, India

A. Kar
IIT Delhi, New Delhi, India

© Springer Nature Switzerland AG 2020
J. Hemanth et al. (eds.), *Data Visualization and Knowledge Engineering*,
Lecture Notes on Data Engineering and Communications Technologies 32,
https://doi.org/10.1007/978-3-030-25797-2_2

1 Recommendation System

As growing information on the web, makes the user confusing in order to choose any appropriate product or an item from the e-commerce websites. To provide the relevant options to the user among the vast number, the concept recommendation system came into scenario. Recommendation system is the application which provides the most suitable suggestions to the user on the basis of his past ratings, user profile, similar users.

There are various steps included in the building of the Recommendation systems:

i.   Database can contains varieties of data like ratings, features of products, user info, ranking of products, user's interest etc.
ii. Various filtering procedures can be used like content based, hybrid, collaborative, context aware.
iii. The techniques like Bayesian networks, neural networks, genetic algorithms, probabilistic methods, K-NN algorithm can be applied.
iv. Then we see the performance of the system in terms of memory and speed.
v.   After that top recommendations are displayed to the user.

1.1 Various Classes of Recommendation Techniques

According to the author [1], there are six different classes of the recommendation system as shown in Fig. 1.

Knowledge Based Recommender System

Recommender system is an application that advices the best course to the user. It strains the information present on the website about the user preferences. On the basis of that the system recommends the choices to the user. Knowledge engineering can be applied on the recommendation system's knowledge base to generate relevant information according to the user's perspective.

As there are number of products in the market, lot of information about them exist. Knowledge of the product deviated frequently. Hence there must be a knowledge base development which upkeep procedures efficiently. Knowledge engineering is a task included in the artificial intelligence that is used to develop the knowledge base systems. These kinds of systems are used to generate knowledge from distinct sources and solved real world problems. Knowledge based recommender system recommends the items by auto generating or non-automatic decision rules. In this system both domain knowledge about the products and user are considered for relevant recommendation.

Fig. 1 Types of
recommendation system

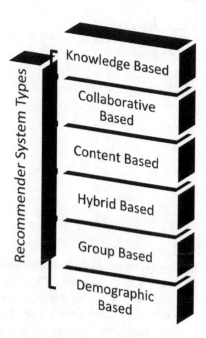

Content Based filtering Technique

In Content Based filtering technique, the main focus is on the features of the items or the products i.e. the content. On the basis of this content the predictions are made. User profiles are analyzed by looking the attributes of the items chosen by that user in the past. In case of document recommendation, content based filtering performance is exceptional. Distinguish models are used to evaluate the similarity among the documents to extract valuable recommendations. Models like Vector space Model (term frequency Inverse document frequency), Probabilistic models (naive Bayes, decision trees, neural network) could be used to make recommendations through learning with machine learning approaches or statistical approaches.

Examples of content based recommender system are Citeseer, NewsWeeder, NewsDude, InfoFinder etc.

Collaborative Filtering Technique

Collaborative Filtering is a prediction technique which works on the database of the user's previously selected preferences of the products. This database is a user item matrix which clutches all the relationships of the user with the item. This works by computing the similarity between the user's choice and the item and then delivers the recommendation [2]. Such users maintains a locality and then the users get the recommendations of different items which he has not rated before. This recommendation is depend upon the other users present in the locality.

For Example, shown in Fig. 2 if there is a user A and there is a locality that is maintained by the user A. This locality (shaded area) is formed by the similar type

Fig. 2 How similarity calculates on the basis of neighborhood

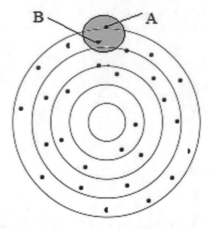

of users who are having somehow similar preferences. There is another user, user B who exists in that locality. Then the user A will get the recommendations based on the user B for the products that user A has not rated before. Actually this happens by evaluating the similarity index between user A and user B preferences.

Learning Transfer using collaborative filtering is divided into two categories, one is Memory-based and other is Model-based as shown in Fig. 3.

Memory based Technique

User's previous choice is important in memory based recommendation. Ratings helps to find the other users exist in the neighborhood. This can be achieved in two ways. First is user-based and second is Item-Based.

Fig. 3 Collaborative filtering techniques

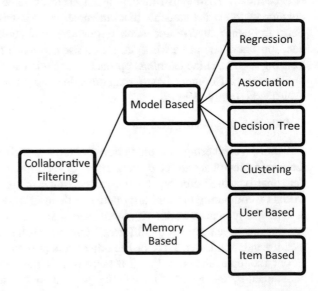

In User-based, evaluation of similarity index among the users is taken into consideration. In Item-based, predictions are formed by evaluating the similarity between the items.

Among all the similarity measures which calculates the similarity of the item with the user, the most common are Pearson co-relation coefficient and cosine based. Pearson co-relation coefficient is used to evaluate that how much the two variables (items) are related to each other. Cosine similarity is based on the linear algebra approach. It belongs to the vector space model. It calculates similarity among multi-dimensional vector which depends upon the angle formed between the vectors.

Model Based Technique

These techniques are based on the various machine learning and the data mining algorithms. In model based techniques, large set of rating collection is used to train the model which gives prediction according to the model designed. The accuracy of the model based recommender system is better than the memory based. There are some algorithms that are used in model based techniques.

(a) Association Rule: In this relationship rules are extracted between the two set of items and based upon the rules predictions are made.
(b) Clustering: In this, useful clusters are formed from the data. Users belong to the same category will belong to the same cluster. Choices of the different users in the particular group can be averaged for the recommendation of the user. Most popular clustering algorithms are K-means clustering and self organizing maps. In k-Means clustering, K clusters are formed with the data depending upon the features provided.
Self organizing maps works on unlabeled data. There is no need to know the class of a particular cluster to work on. Feature map is formed in order to detect the inherent features from the problem. It is used to map the multi-dimensional data to two dimensional data.
(c) Decision Tree Algorithms/CART model: Classification and Regression tree (CART) is another name given by the author Leo Breiman instead of decision trees. These algorithms are meant for classification and regression predictive modeling. Decision tree is a model like a tree structure in which every node is an attribute and every branch is a decision rule. The leaf nodes are the different classes. If the decision trees are trained with high superiority data then upshots into high prediction model with greater proficiency. They are also flexible in handling the data that is either categorical or real valued with missing feature values. Usually this algorithm is known as decision trees but for the languages like R, it has been referred as CART. With the help of CART, the algorithms like bagged decision trees, random forest and boosted decision trees can be founded. CART model is represented in the form of binary tree in which each root node represents the input variable and the leaf node represents the output variable [3].

Examples of collaborative based recommender system are Ringo, Bellcore video recommender, Tapestry, GroupLens, Amazon, Omnisuggest, LensKit, NetFlix etc.

Hybrid filtering

Hybrid filtering is the permutation and combination of content based and Collaborative Filtering techniques in order to prevent the final version of the recommendation systems from the flaws obtained by using the techniques (Content based and Collaborative Filtering).

Demographic Based

Demographic based Recommender systems are the systems that recommends the items or the product on the basis of the personal user attributes like the age, gender etc. The main objective of this system is to track out that which kind of people prefers the products.

Group Based

This type of recommenders systems mostly rely on the user's friend circle. Researches show that most of the user selects the products which are recommended by his/her friends as compared to arbitrary similar users [4]. This brings to the development of the social recommender systems. It is based on the ratings provide by the user's friends.

Comparison of the four filtering techniques is given in the Table 1.

1.2   Limitations of Recommender System

In terms of Content based Filtering

- Limited content analysis
  Extracting valuable information from the content like images, videos, audio, text is very cumbersome task. The problem of filtering out the information from the multimedia content reduces the quality of recommendation.
- Overspecialization
  In this the user gets the recommendation of very similar items but they don't get suggestions of the unknown items. For example there are items i1, i2 ... i10. These items belong to set R1. Then a new item i11 enters into the system. The user will get the suggestions only regarding the items that belongs to R1. The item i11 will not be recommended to the user.

In terms of Collaborative Filtering

- Cold start problem
  There are tri situations for which the cold start problem [21] occurs namely for the new user [22], new item and the new band of users [23]. When a new user enters into the world of ecommerce, the application is not having much information about the user. So the recommendation is not able to provide much of the good recommendations or suggestions. This is the same case with the new item [5, 24].

Table 1 Comparison of filtering techniques

Content based filtering	Collaborative filtering	Demographic filtering	Hybrid technique
• Make recommendations on the user's previous choice • Extract the content from the products or the items for which the recommendation is required • Similarity is evaluated between the items, the user has already visited, viewed or ranked positively • Limitations: first, filtering depends on the item content. Second, content may contain relevant and irrelevant information of the product so matching of keywords can choose the irrelevant information irrespective of the context of the user's choice • Third, deep knowledge of the items attributes is required	• This filtering technique considers the rating which the user gives about the items or products • Main approach used in this type of filtering is K-nearest neighbor [5–7] • This makes prediction according to the similarity between the user's interests with other users • Applications like Ringo, recommends music albums and artists [8], decision-theoretic interactive video advisor [9]	• This type of filtering considers the user's personal attributes like sex, age, country, date of birth, education etc. [10] • Clusters the users depend upon the above attributes and recommend the items accordingly [11] • In this new user recommendation is shown by first selecting the category of the user and then assigning the choices based on the other users of that particular category • Less use of this demographic filtering as it processes the personal data [12] • Can't be used in personal search and education • Less accurate as compare to other techniques [13]	• Hybrid filtering [14, 15] is the combination of either collaborative filtering and demographic filtering [13] or collaborative filtering with content based filtering [16, 17] • Techniques are combined to fight with the flaws of the joining schemes • Results are evaluated by using probabilistic models (Bayesian network, genetic algorithm [18, 19], fuzzy system, clustering, and neural network) or hybrid • handle the flaws of the cold start • Drawbacks [20]: firstly, lack of contextual information to predict the user choice in domain like education is less accurate. Secondly, do not include multiple criteria rating

The new community or the band issue emerged when the new recommendation system started, as sufficient amount of ratings are required to make the appropriate suggestions.

• Sparsity problem
  Sparsity issue occurs when the ratings provided to the item are very less as compared to the number of suggestions is required to predict the items. For example,

if a song is rated by very few users then that song will be less recommended by the RS to the other users although the rating was good.

1.3 Why Recommendation System Became so Popular

The drivers of popularity index of any recommendation system have been depicted as

- Rise the rate of product selling
- Selling of non-popular items also
- Increase the user satisfaction
- Increase the user fidelity
- Understand the user demand.

1.4 Applications of Recommendation System

Recommendation systems can be implemented in different domains. There are number of applications of Recommendation systems in distinguished areas like:

- E-Commerce: There are number of e-commerce websites which used to sell their products on internet. Now days, many users used to buy items of their choice by using these websites. By using the recommendation concept on the websites, displays number of suggestions to the user which results in raising the sale of the products like books, clothes, shoes, hardware devices etc.
- Entertainment: In this case the recommendations of music, movies, games, videos etc.
- Content: In this the recommendations of documents, newspapers, webpages etc.
- Social: suggestions of friends on the social networks, LinkedIn updates, Facebook feeds, tweets etc.
- Services: recommendations of tour and travel services, matrimonial services etc.

1.5 Real-Time Examples of Recommender System

For real time examples see Table 2.

Table 2 Examples and their description

S. No.	Example	Description
1	eBay	In this the recommendation works with the help of feedbacks and the ratings provide by the vendors and the consumers
2	Jester	This is the recommendation system which recommends jokes
3	Last.fm	This is music recommendation system
4	Google News	This recommends the Google news on the basis of the interest of the user in a particular area
5	Facebook	In this the recommendation system is used to provide the new friend suggestions to the user
6	CDNOW	This makes the user to set up their own list of songs on the basis of the artist and albums he likes. Recommendations are on the basis of the list he own and the ratings of the chosen items
7	Pandora	This is an application which is used for music listening and podcast discovery for 70 million users every month with its own Music Genome Project and Podcast Genome Project. Pandora is a subordinate of Sirius XM Holdings Inc. Both Pandora and Sirius jointly have come in the queue of the world's largest audio entertainment company
8	Levis	In this, the user selects the clothes by providing his/her gender. Then the recommendations will be based on the rating, user gives
9	Moviefinder.com	This lets the user to find the match of a movie of similar cast, artist, and genre. Then the system will recommend the list of movies as a result of prior choice
10	Reel.com	Recommends the movie match on the basis of the previous choice of the user
11	GroupLens	GroupLens has started with his original article of USENET. Now there are various projects (MovieLens, LensKit, Cyclopath, and BookLens) that come under GroupLens
12	Amazon.com	Formerly founded as a Book Vendor, trade expanded with many forms of products. Recommendations are given on the basis of user behavior, ratings, browsing activity
13	Netflix movie	Netflix was founded in 1990 and it is an online video shop. Currently it has been spread in over 190 countries and is having more than 100 million subscriptions. This provides the recommendations on the basis of collaborative filtering techniques
14	IMDb	IMDb is launched on internet in 1990 and now it is a subsidiary of Amazon since 1998. This is internet Movie database which stores total of 250 million data items which includes more than 4 million films and T.V. shows and 8 million celebrities
15	YouTube	The video recommender of the YouTube in 2005. Everyday more than 1 billion videos are viewed by over 1 million users
16	Trip advisor	This is the largest website of travel in the whole world having 702 million reviews. This covers around 8 million accommodations, airlines, and restaurants. Trip advisor suggest the best options to stay, flights, have food

2 Literature Survey

Recommendation system plays a vital role in filtering out the valuable options from the large amount of products available on the ecommerce websites. There are number of articles/research papers related to recommendation system that are published in the continuous years as shown in Fig. 4.

It has been observed from Fig. 5 that there is lot of research has been done on the collaborative filtering technique as compared to content based and hybrid filtering approaches.

There are number of algorithms that can be used to implement the recommendation systems. Recommendation systems includes like multimedia based, cloud based, context aware based, IoT based as shown in Fig. 6.

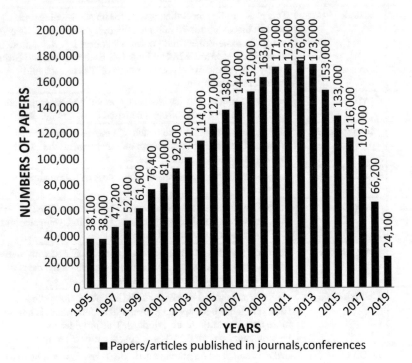

Fig. 4 No. of papers/articles published in journals, conferences on recommendation system

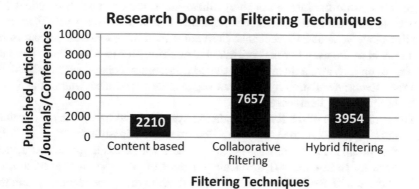

Fig. 5 Graph shows research being done on different filtering techniques

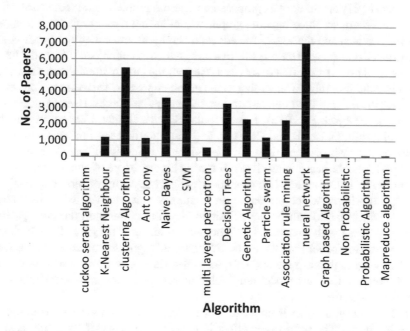

Fig. 6 No. of papers published corresponding to the various algorithms used for implementing the recommendation system

3 Recommendation Systems with Internet of Things

Internet Of Things [25] is an interconnected web of devices based on communicated protocols. IoT is a result of communication of three things: objects, semantics and internet. IoT sensors converse with the help of protocols namely 5G, LORA, Bluetooth Low Energy (BLE) and Zigbee.

To incorporate the internet of things there are some distinguish practices like Near Field Communication, Machine to Machine Communication, Radio Frequency identification, Vehicular to Vehicular Communication are used with respect to the requirement of the user. For making valuable decisions for making choice from the number of options available in front of the user, Recommendation systems are came into role. Recommendation Systems can suggest the items according to the choice of the user by using internet of things.

The performance of the Recommendation system based on IOT is measured by using three well known measures: precision, recall, F-Measure. In paper [26], the researcher proposed a model which includes users which has a profile and devices (IOT devices). In this the IOT devices constitutes of some services and the users will make use of these IOT devices. He found that the recommendation system that provides the suggestions based on popular devices shows the minimum precision value and in contrast, based on popular services shows high precision.

In paper [27] the author has proposed a tripartite graph based model for IOT. In this model there are three bipartite graphs which have a ternary relation among the objects, services and the users. The edges of the graph are called the hyper-edges. This graph develops with the scrutiny of users possession of objects, object deploy by services and user registration to services. Further, the author streamlined the service recommendation problem according to the graph approach. There are various existing algorithms like Most-Popular-Service, Most popular Service User, Most Popular Service user Object, ServRank, Most popular Service Object, User based Collaborative Filtering, Object Based Collaborative Filtering that are implemented by the author on the basis of graph model. He found that the ServRank recommender algorithm performs the best among all as it is having highest score of precision and recall. After this the algorithm Most Popular service object performs better. This concludes that the algorithms which are based on frequency perform well when they select objects. Object based collaborative filtering algorithm comes on the third position when compared with user based collaborative filtering. This gives the conclusion that services are better than the users. Graph based algorithm elucidate the sparsity problem with high accuracy and less execution time. The limitation of this graph based model is that this model is not able to accommodate with localization and mobility of sensors.

The bandwagon effect is important for the recommendation systems to recommend the items. The bandwagon effect is actually a process of spreading positive information regarding the objects on internet. As a result of this effect leads to the rise of recommendation. Sang-Min Choi and others researchers in [28] proposed that how a bandwagon effect can be used in the field of IoT through the recommendation system. GroupLens database is used against the model. The test results show that the famous movies chosen by large number of users is same as the small number of users.

The Current trends and its related search corresponds to IoT is given in [29].

Goswami et al. reported that employing fog computing smart network controls is achieved that distribute video-on demand content due to low latency, with dynamic load balancing thus creates a blue-print for building IoT based recommender system [30].

Existing applications of Recommendation System in the area of IoT

- Health Monitoring: In this [31], the user gets help to look out his daily routine in order to improve his health condition by knowing about the estimating devices. This requires several IoT devices which can deal with related applications to monitor health conditions. A user can use various data visualization tools and can handle his data on cloud and also can share on social web. Wearable devices (activity trackers) and medical sensors automatically connect with the cloud and can upload the data as necessary. User can access his data by using web applications.
- Wildlife animal Monitoring: In this the estimating and data collection devices are set according to the area needs to be monitored to look for wildlife.
- Smart Homes: The recommendation technology helps to improve the overall performance of the devices installed at home.
- Retail services: In order to improve the experience of the buyer in the physical store, recommendation system check that which type of discounts can be shown to the customer.
- Large Scale Sports events: In this, the events like marathon and triathlons, participants can be easily located.

4 Multimedia Recommendation System

As enormous amount of content is available on the web, therefore it is very difficult to choose the useful information from it. The recommendation systems are there to help the users to select the best among the number of options available on the web. As along with text, there is availability of videos, images, movies, songs etc. on the web hence multimedia recommendation systems are there to help the users. Multimedia recommendation system is the application which recommends the multimedia content like images, videos, songs, movies etc. In the field of entertainment there are number of recommendation system like image recommendation system, movie recommendation system, music recommendation systems.

- Image Recommendation System: when a user want to buy an item from the websites like Amazon, PayTm, Flipcart etc. where there are number of images listed corresponding to the particular item. Then there should be a system that helps the user to choose the relevant item of his choice that he actually wants. Image recommendation system is the application that recommends the images of the similar product or suggesting the alternative items. For example, if the user wants to buy the shoes then the system can recommend him the socks images. Similarly jeans can be suggested along with the shirts or t-shirts. In [32] the author develops a system that recommends the image relationships of the various clothes and accessories images. In this the user can evaluate the various combinations of clothes that will make the perfect match.

- Movie Recommendation System: This recommender system is the application that recommends the movies on the basis of rating that user has given in the past or after he had seen the movie. In [33] the author proposes a network centric framework which evaluates the static as well as dynamic characteristics of recommendation systems. The dynamic properties include the polarization and segregation of information between the numbers of users. This model has been implemented on various movie databases like Google Play, IMDb and Netflix. By analyzing these movie datasets, he found that the user will bear the less polarizing and the segregating regarding different movie genres, when compared with other two databases. On the other side, Netflix and Google Play produce more polarization and segregation.
- Music Recommendation: There are number of applications of music recommendation systems. For example, in 2005 the application called MusicSurfer [34] was developed which was content based music recommendation system. It was the automated application that extracts the information regarding the equipment used, beat and harmony from the audio signals of the music. It has been implemented on Magnatune dataset. This application was efficient and flexible enough as it does not required meta-information and ratings given by the users. The author Oscar Celma represents the Foafing the Music system which uses the concept of friend and rich site summary of vocabularies for recommending music depending upon the interest of the user. All the music related information was accumulated by recommendation system feeds depends on the user's profile [35]. Han-Saem Park, the author presented a context aware music recommendation system which recommends the music by following the fuzzy systems and Bayesian network models [36].

4.1 Different Domains of Multimedia Recommendation System

There are different areas in which the multimedia recommendation systems belong as shown in Fig. 7.

Content based multimedia recommendation system

In content based multimedia recommender system the content is the any single multimedia item or the combination of two or more items. For example, the movie is the combination of audio, video, and text (optional), video is the combination of number of images and include sound also, audio is may be the combined activity of sound and text (lyrics). These all example (may be more) are act as a content in CB

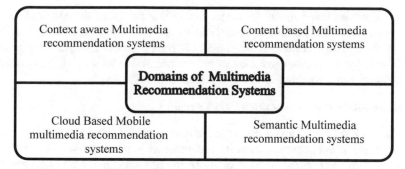

Fig. 7 Different domains of multimedia recommendation systems

Multimedia recommendation system. In [37] the authors has proposed a novel content based video recommender system which filters out the low level visual features (color, lightening, motion) from the videos and suggesting the recommendation on the basis of personalized recommendation without depending upon the high level semantic features (cast, genres, reviews). This leads to the high accuracy of video recommendation. In this they did not consider the algorithm latent semantic analysis which could be applied on the low level features. There are some limitations in this paper which are focused by [38]. The researcher applied various video summarization models based on visual features of MovieLens dataset. This shows that by representing shot features, improves the recommendation quality a lot. In [39], the author applies the deep learning technique and MPEG-7 visual descriptors on the movie recommendation system based upon the visual features. This has been applied on 20M MovieLens dataset. The approach applied leads to the recommendations regardless of high level semantic features like genres and tags.

Cloud Based Mobile multimedia recommendation systems

As the smart phone users are increasing enormously, the multimedia data is also exchanging at a higher rate. This leads to increase in the growth of multimedia content on the web. A multimedia recommendation system and transmission system based on the cloud platform was proposed in which the transmission security was confirmed. All the content of movies was stored on the hybrid cloud platform. For secure transmission this model implemented the encryption and decryption schemes and digital signatures. This leads to the increase in the flexibility of users while exchanging the content [40]. Mobile users face lot of issues while searching for multimedia on the web due to availability of massive content on the web. Therefore a cloud based mobile recommendation system was proposed which was developed by considering the user behavior. Data was stored on the Hadoop platform. This model shows the high values of precision, recall and lower response delay [41].

Context aware Multimedia recommendation systems

The recommendation system, that deals with data of a precise context and delivers recommendations on the basis of it. In [42], the author presented a context aware movie recommendation model for smart phones that considers the context like user preference, network capability and situation to a device. They made a N2M model that follows the hybrid recommendation approach (content based, Bayesian approach, rule-based approach) to suggest the recommendations. He follows MPEG-7 schema for the multimedia content and OWL for context information.

Semantic Multimedia recommendation systems

Cai-Nicolas Ziegler, deployed the imperfections of recommender system by implementing the concept of semantics in the recommendation system [43]. To overcome the flaws of scarcity, semantic concept has been introduced in recommender system as a result of which far better precision results came [44]. Semantic based multimedia recommender system can also be used to browse the multimedia gallery [45].

5 Analyzation Steps for Recommendation System

Progressive steps of analyzing data for recommendation system are shown in Fig. 8.

5.1 Data Pre-processing

This is the preliminary step of creating the recommendation system.

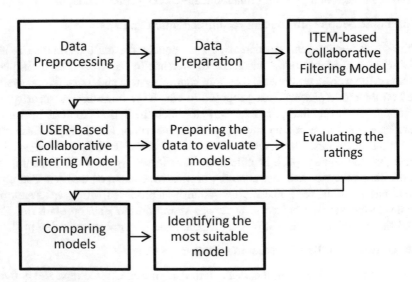

Fig. 8 Steps for analyzing the recommendation system

Extract a list of genres

A list of genres (Action, Adventure, Animation, Children, Comedy, Crime, Documentary, Drama, Fantasy, Film-Noir, Horror, Musical, Mystery, Romance, Sci-Fi, Thriller, and War) for each movie is created so that the user can search accordingly.

Create a matrix to search for a movie by genre

A movie can belong to more than one genre. Hence, a search matrix is created so that the user can search easily according to genres.

Converting ratings matrix in a proper format

To build a recommendation engine, rating matrix is first transformed into sparse matrix.

Exploring Parameters of Recommendation Models

Some of recommendation models include like item based collaborative filtering, User based collaborative filtering, rating based, SVD model. Parameters for these models will be explored.

Exploring Similarity Data

As collaborative filtering relies on the similarity among users or items, hence the similarity will be computed through the use of cosine, Pearson or Jaccard methods.

Further data exploration

The value of ratings is analyzed. Lower value means less rating and vice versa.

Distribution of the ratings

If the value of rating is zero then it seems to be a missing value. Then these values should be removed before visualizing the outcomes.

Number of views of the top movies

Then the number of views of the top movies is evaluated.

Distribution of the average movie rating

Top rated movies are recognized by computing average of ratings.

Heatmap of the rating matrix

Heatmap builds up by visualizing the whole rating matrix. Row represents the user, column shows movie and each cell shows its rating.

Selecting Relevant Users/Movies

Some people are fond of watching movies, so the most relevant users and items must be selected. Number of movies seen per user and number of users per movie must be determined. Then the users and movies matched the particular criteria must be selected.

5.2   Data Preparation

There are three major steps in this process.

Selecting Relevant Data

In order to select the most relevant data, there must be constant value that define the minimum number of users per rated.

Normalizing data

Normalization is required to remove the biasing factor from the rating data. It is done in such a way that average calculation of each user must be zero. Then visualization can be done to see the normalize matrix for high rated movies.

The data is presented in binary form

Recommendation models work on binary data (0's and 1's),
    For keeping the values in the matrix as 0 or 1. There may two situations:

(a) If user gives the rating then value is 1 otherwise 0. In this case rating information is lost.
(b) 1's represent the rating more than the threshold otherwise 0. In this case bad ratings will have no effect.

5.3   Item-Based Collaborative Filtering Model

In this model similarity is measured between two users in terms of similar ratings and other similar users. Then most similar items are to be identified. Further the items are recommended on the basis of similarity to each user's purchases.

Defining training/test sets

Training dataset consists of 80% of total dataset and remaining 20% consist of test dataset.

Building the recommendation model

To make the recommendation model by implementing the IBCF technique, initial step is to evaluate the similarity among the particular number of items and store the count of similar items. There are methods like Cosine (by default) and Pearson which is used to calculate similarity. Finally, a model will be created by using the count of similar items and the method.

Exploring the recommender model

Applying recommender system on the dataset

The algorithm will extract the rated movies for each user. Then ranking will be provided to each similar item in the following manner.

- Extract the user rating of each purchase associated with this item. The rating is used as a weight.
- Extract the similarity of the item with each purchase associated with this item.
- Multiply each weight with the related similarity.
- Then Sum everything up.

Then, the algorithm identifies the top number of recommendations.

It's an eager-learning model, that is, once it's built, it doesn't need to access the initial data. For each item, the model stores the k most similar, so the amount of information is small once the model is built. This is an advantage in the presence of lots of data. In addition, this algorithm is efficient and scalable, so it works well with big rating matrices.

5.4 User-Based Collaborative Filtering Model

In this model, similar users are identified corresponding to the given user. Then top rated items chosen by the similar users are recommended. For new user, the steps are as follows:

1. Evaluate the similarity among the users by using correlation and cosine.
2. Identify the most similar users. The steps are:

   - Find out top k users by using k-nearest neighbor algorithm.
   - Consider the users whose similarity is above a defined threshold.

3. Now, provide the rating to the movies that are rated by similar users. Then the average rating is calculated by using either weighted average rating (similarity as weight) or average rating.
4. Further select the top rated movies.

Building the recommendation system

First check the default arguments of the user based collaborative filtering model and then apply on the training set.

Applying the recommender model on the test set

Then apply the model on the test dataset.

Explore results

Then results will be explored.

Evaluating the Recommender Systems

To compare the performances and choose the most appropriate model of recommendation system, we follow these steps:

- Prepare the data to evaluate performance
- Evaluate the performance of some models
- Choose the best performing models
- Optimize model parameters.

5.5   Preparing the Data to Evaluate Models

There is a requirement of training and test data to evaluate the models. Various techniques are available and describe below.

Splitting the data

In this training and testing data is of 80/20 ratio. For each user in the test set, we need to define how many items to use to generate recommendations. For this, we first check the minimum number of items rated by users to be sure there will be no users with any items to test.

Bootstrapping the data

Another approach to split the data is bootstrapping. The same user can be sampled more than once and, if the training set has the same size as it did earlier, there will be more users in the test set.

Using cross-validation to validate models

The most computationally strong and accurate approach is K-fold cross-validation approach. Through this approach, data can be split into chunks. Now consider each chunk as a test set and evaluate its accuracy. Same will be done with each chunk and at last evaluate the average accuracy.

5.6   Evaluating the Ratings

K-fold approach can be used for evaluation. Initially, the evaluation sets will be re-defined. Then, build item based collaborative filtering (IBCF) model and create a matrix with predicted ratings.

Further the accuracy will be measured corresponds with each user. Most of the RMSEs (Root mean square errors) are in the range of 0.5–1.8. The measures of accuracy are useful to compare the performance of different models on the same data.

Evaluating the recommendations

Alternate way to evaluate the accuracies is to compare the recommendations with the purchases having a positive rating. For this, we can make use of a prebuilt evaluate function in *recommenderlab* library. The function evaluates the recommender performance depending on the number n of items to recommend to each user. We use n as a sequence n = seq (10, 100, 10).

Finally, plotting of the ROC and the precision/recall curves will be there.

If a small percentage of rated movies is recommended, the precision decreases. On the other hand, the higher percentage of rated movies is recommended the higher is the recall.

5.7 Comparing Models

For comparing the models we can do the following:

- Item-based collaborative filtering, using the Cosine as the distance function
- Item-based collaborative filtering, using the Pearson correlation as the distance function
- User-based collaborative filtering, using the Cosine as the distance function
- User-based collaborative filtering, using the Pearson correlation as the distance function
- Random recommendations to have a base line.

5.8 Identifying the Most Suitable Model

Comparison of the models can be done by building a chart displaying their ROC curves and Precision/recall curves.

A good performance index is the area under the curve (AUC), that is, the area under the ROC curve. Even without computing it, the chart shows that the highest is UBCF with cosine distance, so it's the best-performing technique.

Another way of analyzing the success or failures of a movie or its song can be attributed to applications of Natural Language Processing (NLP) computation techniques. Song lyrics analysis leads to analyze certain aspects like we can look out the top words used by the artist in his song, song sentiments, length of the lyrics song, wordcloud of the most popular words in the song and many more.

Fig. 9 Themes sentiment analysis

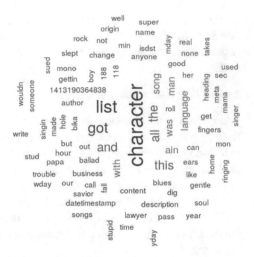

To find the top 15 bigrams in song as shown in Fig. 9 can be executed through sentiment analysis [46]. There are certain sequences of procedure to be followed.

Initially unstructured data gets collected and then text analysis of lyrics by using metrics and spawning word clouds is accomplished. Employing topic modelling techniques such as Latent Dirichlet Allocation LDA techniques. In a nutshell, sentiments i.e. positive and negative both are mapped based on the lyrics and latent theme evolves.

6   Case Study for AI Driven Song Recommender System

Netflix's content recommendation in recent past, Meson is AI interface that automates workflow management and schedule applications. AI gives video recommendations and song that will attract most viewers will help singers to orient their performance that viewers will like most for a longer period of time. For example, let's take the case study of Ms Swati Prasad studied by IntuiComp Terascience Pvt. Ltd. a start-up (http://icterascience.com) into AI based product development R&D based organization.

Using NLP it can be demonstrated how latent talents can get transformed into music career.

AI Engine Prediction by IC Terascience of Swati is shown in Fig. 10.

Using algorithm for search of "Secret superstar", Swati's Professional career dawned from her childhood times. From the Social media analytics, data can be fetched about his performance in different states.

AI shows that Directors and Producers will notice her talent. This "Secret Superstar" had been rightly identified by Artificial Intelligence. based Company [47, 48].

Fig. 10 AI engine
prediction by IC terascience

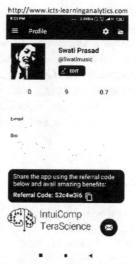

Using social media and Google's search engine, information like her schooling, BE degree, competitions won in BIT, part of the Organizing Core Team in the fest "Pantheon", Bitotsav on Swati's academics was being analysed since 2018 and till now, it had been predicted by AI algorithms that She will sing for few feature films and it happens. Python algorithm can be used to extract tweets or using Google search engine.

Her debut as a Playback Singer in Bollywood movie-"3rd Eye" which is getting released in May 2019. She has sung for few feature films also as predicted by AI algorithms predicted last year.

Employing NLP, prediction of success scope of a budding talents is the essence of Artificial Intelligence application for Entertainment industry

- Initially data collection from Google search engine and social media is to done.
- Secondly, text Analysis of lyrics by using metrics and spawning word clouds is accomplished.
- Thirdly, Sentiment analysis i.e. positive and negative both are looks upon of the lyrics and make inferences.

Predicting which shows will attract users' interest is a key component of the IC TeraScience's model, as user experience personalization becomes more important for the industry, hence, companies are using AI to create personalized services for billions of customers. AI platform can be developed where best talent in entertainment can be identified by using algorithm that identifies best tune and song that everyone that has wider acceptance among audience.

Summary

In this chapter, it has been discussed the various types of recommender system exist, how data scientist can analyze systematically the data to make the recommender system working and achieve the prediction with reasonable accuracy. In this we have also mentioned the case study of the celebrity. It can be demonstrated by employing NLP, Evolutionary Algorithms, Time-Series model, Ensemble methods, Deep Neural Networks etc. nurturing talent can be extrapolated with rate of success prediction in the arena of entertainment industry. AI can imprint huge impact how entertainment features like the music, acting, directing etc. the celebrity can retain and excel to have a notable competitive edge. With the support of AI platforms, best talent among the entertainment industry can be identified and nurtured for years to come.

References

1. Burke R (2007) Hybrid web recommender systems. In: The adaptive web. Springer, Berlin, Heidelberg, pp 377–408
2. Herlocker JL, Konstan JA, Terveen LG, Riedl JT (2004) Evaluating collaborative filtering recommender systems. ACM Trans Inf Syst (TOIS) 22(1):5–53
3. Chipman HA, George EI, McCulloch RE (1998) Bayesian CART model search. J Am Stat Assoc 93(443):935–948
4. Sinha RR, Swearingen K (2001) Comparing recommendations made by online systems and friends. In: DELOS workshop: personalisation and recommender systems in digital libraries, vol 106
5. Adomavicius G, Tuzhilin A (2005) Toward the next generation of recommender systems: a survey of the state-of-the-art and possible extensions. IEEE Trans Knowl Data Eng 6:734–749
6. Bedi P, Sharma R (2012) Trust based recommender system using ant colony for trust computation. Expert Syst Appl 39(1):1183–1190
7. Schafer JB, Frankowski D, Herlocker J, Sen S (2007) Collaborative filtering recommender systems. In: The adaptive web. Springer, Berlin, Heidelberg, pp 291–324
8. Shardanand U, Maes P (1995) Social information filtering: algorithms for automating "word of mouth". In: Proceedings of the SIGCHI conference on Human factors in computing systems. ACM Press/Addison-Wesley Publishing Co., pp 210–217
9. Nguyen H, Haddawy P (1998) DIVA: applying decision theory to collaborative filtering. In: Proceedings of the AAAI workshop on recommender systems
10. Burke R (2007) Hybrid web recommender systems. In: The adaptive web. Springer, Berlin, pp 377–408
11. Aimeur E, Brassard G, Fernandez JM, Onana FS (2006) Privacy-preserving demographic filtering. In: Proceedings of the 2006 ACM symposium on applied computing. ACM, pp 872–878
12. Anand SS, Mobasher B (2003) Intelligent techniques for web personalization. In: Proceedings of the 2003 international conference on intelligent techniques for web personalization. Springer, Berlin, pp 1–36
13. Mobasher B (2007) Data mining for web personalization. In: The adaptive web. Springer, Berlin, pp 90–135
14. Burke R (2002) Hybrid recommender systems: survey and experiments. User Model User-Adapt Interact 12(4):331–370
15. Porcel C, Tejeda-Lorente A, Martínez MA, Herrera-Viedma E (2012) A hybrid recommender system for the selective dissemination of research resources in a technology transfer office. Inf Sci 184(1):1–19
16. Choi K, Yoo D, Kim G, Suh Y (2012) A hybrid online-product recommendation system: Combining implicit rating-based collaborative filtering and sequential pattern analysis. Electron Commer Res Appl 11(4):309–317

17. Barragáns-Martínez AB, Costa-Montenegro E, Burguillo JC, Rey-López M, Mikic-Fonte FA, Peleteiro A (2010) A hybrid content-based and item-based collaborative filtering approach to recommend TV programs enhanced with singular value decomposition. Inf Sci 180(22):4290–4311

18. Gao L, Li C (2008) Hybrid personalized recommended model based on genetic algorithm. In: 4th international conference on wireless communications, networking and mobile computing, 2008. WiCOM'08, pp 1–4. IEEE

19. Ho Y, Fong S, Yan Z (2007) A hybrid GA-based collaborative filtering model for online recommenders. In: Ice-b, pp 200–203

20. Ghazanfar M, Prugel-Bennett A (2010) An improved switching hybrid recommender system using Naive Bayes classifier and collaborative filtering

21. Schafer JB, Frankowski D, Herlocker J, Sen S (2007) Collaborative filtering recommender systems. In: The adaptive web. Springer, Berlin, pp 291–324

22. Rashid AM, Karypis G, Riedl J (2008) Learning preferences of new users in recommender systems: an information theoretic approach. ACM SIGKDD Explor Newsl 10(2):90–100

23. Schein AI, Popescul A, Ungar LH, Pennock DM (2002) Methods and metrics for cold-start recommendations. In: Proceedings of the 25th annual international ACM SIGIR conference on research and development in information retrieval. ACM, pp 253–260

24. Hepp M, Hoffner Y (eds) (2014) E-commerce and web technologies. In: 15th international conference, EC-Web 2014, Munich, Germany, 1–4 Sept 2014, Proceedings, vol 188. Springer

25. Atzori L, Iera A, Morabito G (2010) The internet of things: a survey. Comput Netw 54(15):2787–2805

26. Forouzandeh S, Aghdam AR, Barkhordari M, Fahimi SA, Vayqan MK, Forouzandeh S, Khani EG (2017) Recommender system for users of internet of things (IOT). IJCSNS 17(8):46

27. Mashal I, Alsaryrah O, Chung TY (2016) Performance evaluation of recommendation algorithms on internet of things services. Phys A Stat Mech Appl 451:646–656

28. Choi SM, Lee H, Han YS, Man KL, Chong WK (2015) A recommendation model using the bandwagon effect for e-marketing purposes in IoT. Int J Distrib Sens Netw 11(7):475163

29. Sun Y, Zhang J, Bie R, Yu J (2018) Advancing researches on IoT systems and intelligent applications

30. Goswami D, Chakraborty A, Das P, Gupta A (2017) Fog computing application for smart IoT devices in agile business enterprises. Glob J e-Bus Knowl Manag 8–13(1):1–8

31. Erdeniz SP, Maglogiannis I, Menychtas A, Felfernig A, Tran TNT (2018) Recommender systems for IoT enabled m-Health applications. In: IFIP international conference on artificial intelligence applications and innovations. Springer, Cham, pp 227–237 (2018)

32. McAuley J, Targett C, Shi Q, Van Den Hengel A (2015) Image-based recommendations on styles and substitutes. In: Proceedings of the 38th international ACM SIGIR conference on research and development in information retrieval. ACM, pp 43–52

33. Dash A, Mukherjee A, Ghosh S (2019) A network-centric framework for auditing recommendation systems. arXiv preprint arXiv:1902.02710

34. Cano P, Koppenberger M, Wack N (2005) Content-based music audio recommendation. In: Proceedings of the 13th annual ACM international conference on multimedia. ACM, pp 211–212

35. Celma Ò, Ramírez M, Herrera P (2005) Foafing the music: a music recommendation system based on RSS feeds and user preferences. In: ISMIR, pp 464–467

36. Park HS, Yoo JO, Cho SB (2006) A context-aware music recommendation system using fuzzy bayesian networks with utility theory. In: International conference on fuzzy systems and knowledge discovery. Springer, Berlin, pp 970–979

37. Deldjoo Y, Elahi M, Cremonesi P, Garzotto F, Piazzolla P, Quadrana M (2016) Content-based video recommendation system based on stylistic visual features. J Data Semant 5(2):99–113

38. Deldjoo Y, Cremonesi P, Schedl M, Quadrana M (2017) The effect of different video summarization models on the quality of video recommendation based on low-level visual features. In: Proceedings of the 15th international workshop on content-based multimedia indexing. ACM, p 20

39. Deldjoo Y, Elahi M, Quadrana M, Cremonesi P (2018) Using visual features based on MPEG-7 and deep learning for movie recommendation. Int J Multimed Inf Retr 7(4):207–219
40. Yang J, Wang H, Lv Z, Wei W, Song H, Erol-Kantarci M, Kantarci B, He S (2017) Multimedia recommendation and transmission system based on cloud platform. Futur Gener Comput Syst 70:94–103
41. Mo Y, Chen J, Xie X, Luo C, Yang LT (2014) Cloud-based mobile multimedia recommendation system with user behavior information. IEEE Syst J 8(1):184–193
42. Yu Z, Zhou X, Zhang D, Chin CY, Wang X, Men J (2006) Supporting context-aware media recommendations for smart phones. IEEE Pervasive Comput 5(3):68–75
43. Ziegler CN (2004) Semantic web recommender systems. In: International conference on extending database technology. Springer, Berlin, pp 78–89
44. Thanh-Tai H, Nguyen HH, Thai-Nghe N (2016) A semantic approach in recommender systems. In: International conference on future data and security engineering. Springer, Cham, pp 331–343
45. Albanese M, d'Acierno A, Moscato V, Persia F, Picariello A (2011) A multimedia semantic recommender system for cultural heritage applications. In: 2011 IEEE fifth international conference on semantic computing, pp 403–410. IEEE
46. NLP Coding for Movie and Song Predictors. https://play.google.com/store/apps/details?id= com.lms.android.lms
47. Learning Platform Developed by IntuiComp Terascience Pvt Ltd. http://www.icts-learninganalytics.com
48. Learning platform Developed by IntuiComp Terascience Pvt Ltd. Android application https:// play.google.com/store/apps/details?id=com.lms.android.lms

Image Collection Summarization: Past, Present and Future

Anurag Singh and Deepak Kumar Sharma

Abstract With recent trends in data, it is very evident that more and more of it will be continued to be generated. It is suspected that our limit to provide services to customers will be limited by the type of analysis and knowledge that we can extract from the data. Images constitute a fair share of information in the large form of media that is used for communication. For example text, video, audio to name other few along with their meaningful combinations. While Summarization of videos and events have been of recent interest to computer vision and multimedia research community. Recent advances in the field of optimization especially deep learning have shown significant improvements in video summarization. Image Collection Summarization is an important task that continues to elude because of the inherent challenges and its differences from video summarization. Since the video has a lot of temporal link between the frames that can be exploited using some temporal neural networks like Long Short-Term Memory (LSTM) or Recurrent Neural Networks (RNNs) they prove to be useful in case of designing deep learning based architecture for the event and video summarization. Similarly, for text, it can be acknowledged that a long passage and sentences will have a high-level temporal sequence between them which can be exploited for summarization. While in case of a collection of images there is no temporal sequence between two images to be exploited by the network [14, 24]. This has resulted in the problem being esoteric in nature. To remedy this, the following article plans to bring the challenges in the field of image collection summarization, the need for gold standards in the definition of summarization, datasets and quantitative evaluation metrics based on those datasets and also major papers in the area that have aimed to solve the problem in past.

A. Singh
Division of Information Technology, Netaji Subhas Institute of Technology, New Delhi, India
e-mail: anurags.it@nsit.net.in

D. K. Sharma (✉)
Division of Information Technology, Netaji Subhas University of Technology (Formerly Netaji Subhas Institute of Technology), New Delhi, India
e-mail: dk.sharma1982@yahoo.com

© Springer Nature Switzerland AG 2020
J. Hemanth et al. (eds.), *Data Visualization and Knowledge Engineering*,
Lecture Notes on Data Engineering and Communications Technologies 32,
https://doi.org/10.1007/978-3-030-25797-2_3

Keywords Image corpus summarization · Multimedia · Visualization · Data interpretability · Computer vision · Deep learning · Clustering · Sub-modular functions

1 Introduction

With the recent surge in the ubiquity of the internet, a massive amount of data is being generated. It has become important for businesses and a whole range of other applications to leverage that power of data. Even for purpose of social good, usage of data can provide us effective insights to challenges with mammoth issues that pose a severe threat to mankind. Such as global warming and climate change, poverty and socioeconomic disparity. Now, to derive the valuable insights from such large corpus of datasets, it shall require special efforts and will pose a new set of challenges as the scale of data continues to grow.

Summarization of such large datasets to form a concise representation of the original data, such that it captures all the characteristics within the data and its essence is still maintained. Then when the summary is manageable, valuable analytics and insights can be drawn from them by building mathematical models for prediction and with the help of machine learning techniques. Summarization will also help in an efficient display, carousel and representation of relevant data in multiple industries and will be of value for the current internet and e-commerce industry. Also, with the amount of legal documentation being done in current form summarization is to definitely help people with no background of law to understand the cause and effects of things state in the clauses. It will also help the user augment his understanding of the language tools being used in legal documents to represent or justify situations i.e. rhetoric, issues, claims, supports and double negations. Thus, a more succinct representation helps in settlement of a lot of claims and disputes and also reduces the chances of getting source misconstrued.

1.1 Image and Video Summarization

The point of summarization is to generatively auto summarize the contents of the image or of the video. Therefore there are two types of summarization

- Extractive Summarization: It is the task of summarization with the approach of extracting a representation summary from all the set of images or key-frames within the frames in a video, or in more appropriate words a video skim. Mathematically it can be said as, given a collection of n images $X = (X_1, X_2, ..., X_n)$ we aim to find a summary S such that $S \subset X$ and $|S| < n$, function f such that $\forall s \subseteq X$ $S = \arg\min_{|s|}(max f(s))$ is the most succinct extractive representation of the original data.

– Abstractive Summarization: It is the task of summarization with the approach of extracting information with not extracting direct information but preserving the internal semantic representation of the data. Data may be transformed into a different space in the process of abstractive summarization. In other words, summary may not contain the data from the source but its essence is captured. In text summarization, it could be best explained as similar to how humans paraphrase. In the domains of image and video summarization abstractive summarization does not hold much relevance and extractive summarization holds to be more useful and has more practical applications. Mathematically it can be said as, given a collection of n images $X = (X_1, X_2, ..., X_n)$ we aim to find a summary S such that and $|S| < n$, function f such that $\forall x \subseteq X$ and $s = \sum_x w_i x_i$ $S = \arg\min_{|s|}(max f(s))$ is the most succinct abstractive representation of the original data.

Issues in Definition of Summary There have been multiple attempts in image corpus summarization to deal with the issue of summary evaluation [19], many approaches have been proposed and still the current gold standard lacks approval from researchers as it continues to attract new methods to estimate the efficacy of a summary for every published work. Photo summarization can be very subjective, according to authors in [15] there are a lot of differences within the user preferences of what constitutes as important and relevant. Therefore, user-preferences and intent of task for which the summary is being designed hold a very important place in identifying what should constitute in a summary and can vary across multiple editors.

To totally circumvent this discussion, lot of published work in the direction to solve problem of image corpus summarization considers annotation of summaries and datasets used This gives them a ground truth as a strong baseline to compare with their summarization results. The performance usually involves the discussion of metrics that are well-established such as finding the precision-recall and F-scores for the efficacy of summary.

In [23], authors try to define a different concept for the summarization. It rests on the principle that a summary is a succinct representation of the corpus. Working in fashion similar to compression, it treats summarization and compression as complementary. If a summary is able to reconstruct a close representation of they original corpus in a latent space then it has all the necessary information in it to make it a succinct representation of the corpus. With recent developments in deep learning, in [17, 18] models were suggested to build summarization based on deep learning architectures. In [18] Task specific summarization was introduced where the deep learning architecture learns also about the intent of the particular use of the summary.

2 Literature Review

Image summarization is an important task for several applications such as effective browsing, fast consumption of content also requires the idea of summarization at heart and also in particular efficient navigation in web-data and training of the deep learning based models. It is well known that current state of the art networks for many problems have millions and billions of parameters in their architecture that are tuned using large corpus of data. Even for fine-tuning, deep learning architectures may require large amount of data and training time. Much of that data doesn't contribute to learning and is usually redundant in defining the decision making kernel of the architecture. Thus, summarization finds an interesting purpose in these cases where we have to prototype different models for testing them out and then using the whole dataset to train the network that performs best. Therefore, helping in faster prototyping.

2.1 Similarity-Based Approach

In [16], the authors Simon et al. Use the problem of scene summarization to address the issue of summarization of a collection of images. Their data is prepared by means of selecting the set of canonical images from the web archives for a scene. Authors define canonical views as the set of views of an geometrical object that are able to capture its essence. They try to approach the problem being influenced by the seminal work of Palmer et al. in [13] which tries to define a canonical view of an object. This definition rests to the heart of all the work in [16] and is still being widely researched by the computer vision research community. Authors in [16] try to paraphrase the work of [13] in four different points which are critical to any view of the collection:

1. Given a set of photos, which is most liked view?
2. What would be your choice of view when taking the photograph of the object?
3. What is the view from which object can be most easily recognized?
4. What view will you imagine the object from?

There has been significant research and hence resulting literature on computing canonical views for an object. From perspective of computer vision research, the criteria of Palmer et al. are not applicable for computer vision tasks as they are for a human observer. Therefore, the canonical views definition in computer vision research has focused on identifying the algorithms based on the geometry of object or set of pictures of the object. Taking example of which, in [4] Freeman et al. and in [22] Weinshall et al. have estimated the likelihood of an image view by using things such as: analyzing similar viewing condition producing ranges and also using information from the geometry of the object. Closely relating to the work of authors in [16], authors in [3] using semi definite programming based approach to select a representative set of summary to a give collection of images. Choosing views from the non-canonical views which are actually similar to the set of larger views that are non-canonical, while simultaneously being dissimilar to each other. Thus trying to fulfill the conditions of relevance and diversity in a summary. In [8] Owen and Hall used images with low likelihood as criterion for selecting images for canonical views. Given that the selected set of images remain orthogonal to each other. A computation of the eigen model which is of 10 to 20 dimensions is done, with images from the represented set. Now, using the model the algorithm iteratively extracts the set of images keeping the images in the set orthogonal to each other. Therefore they give the concept of image likelihood where the similarity is measured as an formulation similar to cosines:

$$similarity(V_a, V_b) = \frac{|V_a \cap V_b|}{\sqrt{V_a V_b}} \qquad (1)$$

where $V_i \in \mathbf{V}$ is a view corresponding to a image frame which contains certain set of features that the view of an object must contain. Now when the authors try to cluster their images for a set of canonical views it is needed to maximize an objective function that defines when the clustering has been appropriately able to segregate the differences in the feature vector. Where each vector embedding for an image frame V with its corresponding near canonical view $C_{c(i)} \in C$ where c shall define the mapping of views in image frames to canonical views. There is a need to penalize the solutions that contain all the different possible canonical views as it would make redundancy prevalent in the summary and summaries for a given scene need to be readable as well which is penalized using a hyper-parameter α. Also to emphasize on orthogonality of the differently selected canonical views another term is added to the overall objective function:

$$cost\,function = \sum_{V_i \in V} (V_i.C_{ci}) - \alpha|C| + \beta \sum_{C_i \in C} \sum_{C_{j>i} \in C} (C_i.C_j) \qquad (2)$$

The above equation is designed to especially penalize the heavy similarity between different canonical views and promote orthogonality in the representative scene summary.

2.2 Reconstruction Based Approach

For a given set of images in a corpus, most of the image collection summarization approaches work in the direction of selecting a subset of images from the collection that try to highlight the best and maximum plausible view of the set. Therefore, in hindsight modelling it as a optimization problem. In [23], authors work in this fashion to model the problem of image corpus summarization as an optimization problem. Let n is the total number of images in the corpus and k is the number of optimal images in the summary. Therefore, they define $X \in R^{d \times n}$ as the set of images in the corpus and $D \in R^{d \times k}$ where $k << n$. Then the global optimal number of images in the summary is optimized by the reconstruction error function which is the L2-norm:

$$Reconstruction\,error = \min_{D} \left(\left\| X - f(D) \right\|_2^2 \right) \qquad (3)$$

Here $f(.)$ is the reconstruction function whose purpose is to define a method to convert the set of images selected in the summary to be used to reconstruct the closest replication of the dataset. Authors in [23] define their reconstruction function to be a linear regression model for the summarization task.

The research field of image summarization and state of the art has been influenced using the approach of reconstruction for summarization task. Most of the existing research on summarization is crippled by the bias in evaluation and lacked objectivity. Research done uses the evaluation by means of assigning satisfaction and relevancy score of the user. Although, intent is considered in this case it formulates a weak case for generic summarization for large datasets for which correct evaluation of context, relevancy and succinctness of summary is a challenging task. Formulation of summarization as a reconstruction based optimization task further directs the research using a fresh perspective and enables the new research to take inspiration from methods and techniques used in signal processing and image compression to be used for summarization purposes. Therefore, in addition to the subjectivity with respect to the quantitative evaluation it also introduces reconstruction error as MSE for objective evaluation of the performance of the method. To find the coefficients of the reconstruction function and collection of the summary. Sparse representation of the bases is used and hence, formulation of automatic summarization as an dictionary learning for sparse representations helps the authors to leverage the mathematical tools of linear algebra for solving problems in sparse space representations. Since the objective is defined as an optimization problem it is also plausible to leverage global optimization algorithms for learning the members of the dictionary and also the coefficients of the sparse representations. Therefore, based on the concept of sparse representations the expression of reconstruction error is as follows:

$$\min \sum_{i=1}^{n} \left\| x_i - \sum_{j=1}^{k} d_j \alpha_{ji} \right\|_2^2 \tag{4}$$

Here the images to be the part of the summary consist of $\{d_j | d_j \in D\}$ and the coefficient α_{ji} is the non-negative sparse coefficient weight for the corresponding d_j. In the problem of image summarization $\{d_j | d_j \in D\}$ set consists of the images in the summary which form the succinct representation of the image corpus. The method works on the idea similar to non-negative factorization of a matrix. In fact, if the sparsity coefficient is heavily penalized and then restricted to $||\alpha||_0 = 1$ then the reconstruction will reduce to selection of the single image that can serve as best representation of the given set at current value of error. Which will be equivalent to k-mediods (which is a form of k-means clustering). and is known as an effective method for summarization of the collective datasets. Considering the fact that $|D| << n$ needs to be a constraint in minimization of the given reconstruction loss it is very important to put a check on the sparsity coefficient α for better and more succinct representation. Since the number of bases used in the reconstruction implies the number of images in the summary it is important that we must regulate the same by considering only bases with non zero coefficients. Moreover, the bases must be as diverse as possible within themselves and must contribute to the visual diversity of the summary. Hence the equation becomes:

$$\min_{D,A} \sum_i \left\| x_i - D\alpha_i \right\|_2^2 + \lambda \sum_i \left\| \alpha_i \right\|_0 + \beta \max_{j \neq k} corr(d_j, d_k) \tag{5}$$

Therefore, the problem of image corpus summarization is formulated as an optimization problem in Eq. (5). The optimization of the dictionary D which is a representative set that forms the summary of an image corpus and matrix of corresponding non-negative coefficients α represented using symbol $A = [a_1^T, \ldots, a_n^T] \in R^{1 \times k}$. Here the diversity is formulated using an correlation function between two supposed dictionary vectors. It is calculated by the means of correlation in the two vectors rather than the euclidean distance or the cosine distance. The maximum value of the correlation is robust to the shift and scale, is usually invariant to both of them at the same time. It is calculated as follows:

$$corr(d_i, d_j) = \frac{(d_i - \bar{d}i)(dj - \bar{d}j)}{\sigma_i \sigma_j} \tag{6}$$

Here $\bar{d}$ is the mean of the vector d and σ corresponds to the standard deviation of the vector. The problem of optimization described in Eq. (5) is NP-Hard and therefore falls into the category of intractable problems. Most, of the algorithms that are run to attain a good fit shall fall into the trap of local maxima and hence will not be able to deliver a good fit for the solution or rather give a trivial solution to the problem. The convex optimization problem is the approached via a multiplicative algorithm and the objective is non increasing with the given update rule:

$$A_{t+1} = A_t . * (D^T X) ./ (D^T D A_t + \lambda 1) \tag{7}$$

Here $.*$ is the hadamard product which means element wise matrix multiplication and similarly, $./$ is element wise division. The Dictionary D and the coefficients matrix A are updated in turns thus at any point is saved as the tuple $(D_{optimal}, A_{optimal})$ which keeps $r(D_{optimal}, A_{optimal})$ minimum. It is then after the sufficient number of iterations that the annealing starts and the update terminated when the $r_{optimal}$ is not updated for a fixed hyper-parameter that is, MaxConsecRej(in other words the maximum number of consecutive rejection) times the number of iterations. The objective is then evaluated for summary against several baselines for assessing the efficacy of the summary.

2.3 Task-Specific Summarization Based Approach

There have been few novel attempts to look into the interest within the summary and to evaluate the efficacy of the summary with keeping in mind the intention for its creation. In their paper [19], Ramesh Jain et. al essentially propose a model for summarization of photo albums using multidimensional information both in the form of content and also in the form of context. The idea is to generate a automatic summarization framework which shall help to achieve a task of giving an overview of information present within the dataset. In true sense trying to highlight its regularities and well as its irregularities within the given space constraints. The system should be able to interact with the users and be able to generate summaries with the fundamental intent for which the user wants. To model the intent quantitatively certain set of hyper-parameters are kept that ultimately quantify the summary both in quantitative and qualitative terms. Thus exploring the possibilities of better design of photo navigation systems which shall help in retrieval of similar kind of images to which the user wants. Essentially drilling down on right kind of photos in much quicker time by means of easy navigation as supposed to both content and context unaware search. The three basic properties that are identified for association with an image from a dataset are an *Interest* within the image. *Interest* is inherently an context aware quantification of the image frame. So, it is easier to reason that the image with higher interest must be the one to be included in the summary. Second, there is a concept of the distance between the two image frames $distance(x, y)$ is the distance between the representation of image frames x, y in the representation/concept space. Third, $represent(x, y)$ is used to quantify the set of elements in the two images x, y which are common to both in some representation space x, $y \in \mathcal{R}^n$ where n is the dimension of the representation in the space. There are three effective points for summarization of datasets:

1. **Quality**: The photo summary must serve the underlying interest of the consumer and hence must be built in an very context aware manner. The summary $\mathcal{S}$ is the collection from the images x in dataset and its quality can be quantified as

$$Q(\mathcal{S}) = \sum_{x \in \mathcal{S}} interest(x) \tag{8}$$

2. **Diversity**: The images in the summary must be sufficiently distinct from each other and must be covering some unique element of the image dataset. For the given constraints the summary must not be redundant and should avoid repetitions of the views or the images it needs to be diverse enough in itself. It conveys similar idea to orthogonality as in [16]. In the given paper the idea for diversity is related to the distance in the representation space of the image frames. Therefore in aggregation of the image frames it means that for any two pairs of the image frames in the summary what is the minimum distance between the frames is equivalent to diversity. While this idea of quantifying diversity is limited and too strict because the minimum distance in the pairs may not actually reflect the diversity of the images in summary on average.

$$diversity = \min_{x, y \in \mathcal{S}, x \neq y} distance(x, y) \tag{9}$$

3. **Coverage**: The coverage as a quantifiable quantity is defined as ability to represent succinctly all the information that is contained within the original image corpus.

The measure $Cover(x, X)$ is the actually the sum of the number of images in the corpus i.e. X which are related to the image x. Essentially it tells that what is set of images from the corpus does the image x cover. In order to compute the cover for the summary S selected from the corpus we need to aggregate the covers from all the images present in the summary. An perfectly optimal summary will have coverage of complete image corpus, in other words it will be sufficiently be able to represent the whole corpus without losing out on any valuable frames. This shall also depend on the other restrictions by the user such as size of summary and intent. Nevertheless, an optimal automatic summarization model must be able to give maximum amount of coverage.

$$Cover(x, X) = \sum_{y \in X} |represent(x, y)| \tag{10}$$

$$Cover(S, X) = \sum_{x \in S} Cover(x, X) \tag{11}$$

Similar to Ramesh Jain et al., Subramanyam et al. [18] in their work on summary generation using deep learning models also promote the usage of intent or an idea of task for which the summary is being generated. Their classification of summary generation and its essential properties are different than those posed in work of Ramesh Jain et al. An succinct representation, in other words an optimal summary must contain two properties:

Relevance: The relevance of an image is basically how important it is to understanding of the characteristics of a dataset. And how well it fits in the context of the overall objective related to task for which the summary is designed.

Diversity: The diversity of an image set is basically defined as how much different view or perspective do two different images add to the summary. The summary must contain all the different set of characteristics that it is supposed to capture.

Redundancy: The amount of images in the summary must be controlled and there should not be any repetition of images in the summary. The length of the summary also dictates which of the other properties are most reluctant in the summary. Often small set of summary images might not contain fair share of representation for the all the images present in the dataset, in such cases it becomes important to know which properties to forgo in the summary in order of it to be the most optimal representation of the given dataset in given constraints. Also there is a knee-break point in every dataset where certain set of images can be removed without hurting much on the above two properties of the dataset, in other words the representation doesn't contain any redundancy then,

Task-specific: task specificity means that images are chosen with a particular task in mind. The change of task can result in changes of images in the summary. More often than not it is important to know what contents are relevant for a given context and would serve best purpose. The idea of task specificity is achieved by knowing the

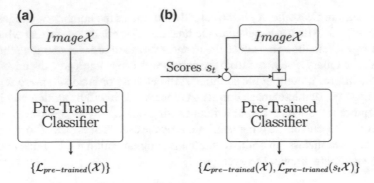

Fig. 1 The **a** and **b** introduce two different variants of the loss in the task specific manner. Here the difference is that the first loss uses score in its formulation making it a linear loss with respect to the scores. Where as second loss is where task specific loss pipeline first gives a output for image and then for the score times image. This induces non linearity in the formulation and hence can capture more complex decision boundaries

performance and relevance of a image for a given task by using a pre-trained classifier on it. Then using that information the authors in Subramanyam et al. designed an $\mathcal{L}_{task-specific}$ which is the task specific loss which tells the importance of a given image in the summary as its minimum for a particular score will help select/reject the image frame. The two equations which the authors experiment with for the task specific loss are as follows, here β is a hyper-parameter to control the effectiveness of loss in over all objective function. And s is defined as the individual score assigned to each image about its importance in summary (Fig. 1):

$$\mathcal{L}_{task-specific} = \frac{(1-s)\mathcal{L}_{pre-trained}(\mathcal{X})}{\beta}$$
$$\mathcal{L}_{task-specific} = \frac{||\mathcal{L}_{pre-trained}(\mathcal{X}) - \mathcal{L}_{pre-trained}(s\mathcal{X})||}{\beta} \qquad (12)$$

Now when the summary is generated with best relevance and diversity possible for a particular task it can be said as the optimal summary for the use in that particular task. Moreover, authors in Subramanyam et al. [18] mention how training of deep neural network based architectures proves to be cumbersome and time taking task. It is even hard to fine tune models as it takes heavy compute and lots of data for even simple tasks. As our models become more intelligent and more well-equipped to generalize with lesser amounts of data it is also important for us to be intelligently be able to down-sample the datasets such that both its majorities and irregularities remain preserved. It must not lose its shape while down-sampling in a concept space. Which in other words means that it should remain topologically invariant. Thus the evaluation against a task is also proposed as a metric for quantitative evaluation of the metrics. It is important in same sense that how well can the given set of images perform for the particular task does tell us something about the quality of images

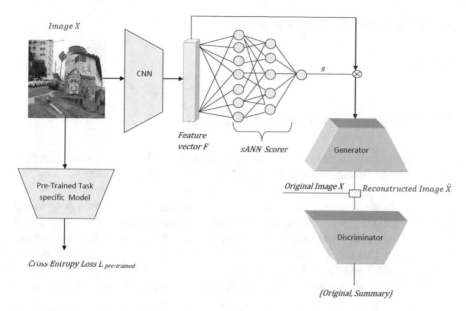

Fig. 2 The above diagram shows how a pre-trained classifier can be integrated with a full fledged deep learning based model for task of summarization. The trained classifier gets same set of images as inputs and also images multiplied with their respective scores from the classifier. Finally the objective loss is determined based on the training of the pre-trained classifier for the given task for which it was trained

in the summary. Moreover, it is also relevant as a evaluation criteria because well for some kind of tasks certain images must prove to be hard examples which must be there in the training of that task. As they are the data points which marginally form the decision boundary and support the kernel of the classifier in separating the different hyper planes associated with different classes. It is important to formulate the task-specific loss in a manner that is a primal dual optimization. Where it is important to maintain both relevance and diversity where each of them has opposite characteristics to ensure in the summary. For example in case of Subramanyam et al. [18] where they introduce the idea of interest or task specificity in the deep neural network architecture. A diverse set of images is a set of images for a given task that will happen to be independent of the feature space and add new things to the space whereas the relevant set of images will be the images that help fix the boundary of the classifier which happens to be an outlier. More often than not the two set of images may not be the same. Hence regularization is needed for the task specific loss to control the number of outliers in the summary and hence ensuring both relevance and diversity. An overall architecture for task specific summarization is discussed in the Fig. 2.

3   Techniques in Summarization

See Fig. 3.

3.1   Deep Learning Based Architectures

Image corpus Summarization has comparatively received less attention as compared
to Video based Summarization model [12, 18]. In recent few attempts to describe
the task of the image corpus summarization as an objective that can be used to train
a neural network. It is following idea that is used to train a scorer based deep neural
network in an end to end fashion. The idea is to generate features of the dataset image
frames using transfer learning from a fined tuned Convolutional Neural Network
on ImageNet Dataset. It is relevant features which are then propogated through
a scorer which labels each feature with its importance in the summary using the
score $s \in [0, 1]$. This score is then multiplied to the feature vector X to embedded

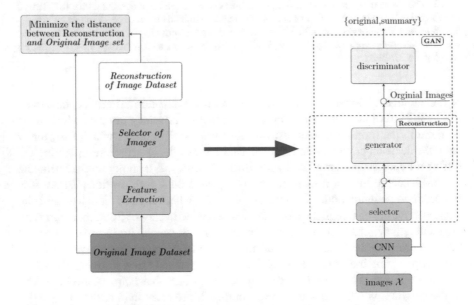

Fig. 3 The above figure describes the essence of building and summarizing using deep learning
based models. It serves as a sort of teaser to building models. The general idea served in these
models is that model iteratively compresses the images into feature vectors and extracts a latent
representation of the features. This representation is fine tuned by reconstructing the copy of original dataset from the representation. The difference in original and reconstructed dataset is the
loss of information experienced by the network. Optimization is done to learn and minimize this
loss. Afterwards, this representation is used to understand the data and pick summary by various
techniques

Fig. 4 In the following figure a deep learning model is presented by the authors in Subramanyam et al. [18] as they try to model a scorer S which will output a soft-margin score corresponding to the probability of an frame being part of the summary

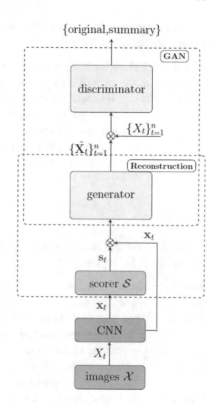

the relevant information of what scorer neural network thinks is importance of the feature in summary. The feature is then reconstructed using the generator into the complete picture of the image $\hat{X}$. Now this reconstruction is compared to the original image frame X to find out the reconstruction error in the set. This loss is used as one of the regulation losses in the deep learning based summarization model for the scorer neural network.

Length Regularization Also, the number of images that need to be regulated to be within the summary are to be chosen since the summary is strictly less than the total number of images in the dataset. In fact, the score matrix over the number of epochs must be sparse enough. To regulate this a general length regularization loss is needed (Fig. 4).

$$\mathcal{L}_{length-Reg} = ||\frac{1}{N}\sum_{i=1}^{N} s_i - \sigma||_2 \tag{13}$$

where σ is a hyper hyper parameter that controls the length of the summary, over course of the epochs the number of images in the summary gradually become limited to the number σn. In other words, the images are selected into summary with greater confidence where most of the scores are pushed close to 1 when the are selected in

the summary and the number of images in the summary that are not to be selected
are having scores close to 0.

Key-Frame Regularization There are lot of key frames in the image corpus that are
important to the overall summarization task both in the generalized summarization
and also for any task specific summarization. The key frames are expected to have
high scores from the scorer network since they are well able to reconstruct the
original dataset and make most succinct representation of the corpus. Key-frame
regularization is only possible to make when there are annotations for every frame
based on its relevancy and diversity.

$$\mathcal{L}_{Key-frame} = \frac{1}{N} \sum_i cross_entropy(s_i, \hat{s}_i) \tag{14}$$

where N is the number of images in the dataset and $\hat{s}$ is the annotated value of the
ground truth for a particular frame.

Repelling Regularization It is also important to repell the similar kind of frames in a
corpus. To ensure that the losses in the subspace repell similar kind of images for them
to have a hyper plane separator classify them as different or to punish the classifier
for selecting lot of images that are similar to each other repelling regularization is
essential to ensure the diversity in the summarization task [26]. There general idea
is to first construct a similarity matrix L which is a $N \times N$ matrix where each entry
$L_{i,j}$ denotes the similarity between the images frames X_i and X_j. It is trivial to show
that the matrix is symmetric and is expected to be sparse enough.

$$\mathcal{L}_{Repelling-Regularization} = \frac{1}{M(M-1)} \sum_i \sum_{j \neq i} \frac{(s_i X_i)^T (s_j X_j)}{||s_i X_i|| ||s_j X_j||} \tag{15}$$

Thus, repelling regularization can ensure that the frames with high visual diversity
are taken into account based on the fact that if selection is made such that the scores
s corresponding to set of images that are unique in the space plane then they would
be easily separated with the rest of the features from the frames as the product of the
feature vector and low scores will be a zero-vector.

3.2 Clustering

Similar, to work done in deep neural network based architecture methods clustering is also employed to select the subset of images from the corpus. Clustering has
been fairly old and established method of summarization for different media. Several attempts have been made to cluster the data and then find out redundancies in
the set [1, 2]. Also borrowing implementations from the image compression ideas,
summarization and compression along with saliency detection can be said to be very
much different forms of a similar problem of identifying closed form representa-

tion of the data. This has also resulted in the techniques of topological data analysis which is shape and form invariant. This helps in identifying the key data points that would make that data behave in same manner in that feature space. In other words finding the pivotal points in that data space cluster to ensure that the selected points still ensure similar essence of data by having same topological shape as the original corpus. Thus clustering is core to the idea of summarization. It is very hard to image simultaneous summarization of images without clustering the points in the feature space on the basis of a metric that is being learned to identify the distances between the features given our objective. Also, it is common to see that the data is decomposed or transformed to another subspace which acts as an auxiliary subspace for the data. The parameters of which are learned from the convex optimization formulation. Now in that subspace the features that are similar can be clustered together to obtain points similar to each other. Several ensured diversity selecting means can be incorporated considering that the distances between the similar kind of images turn out to be euclidean after the transformation is applied to the data. In several cases all the transformations are complex or too computationally expensive to run on the traditional form of the data. Therefore there is a need to shun a lot of information and obtain a feature vector representation of that data. Several feature selection methods can be employed for the same task. Since deep neural nets are good given the large amount of data and computation power when a existing neural net can be fine tuned using the transfer learning from some dataset. One of such work was done by the authors Sharma et al. [17] in which authors use convolution neural network based architecture and also auto-encoders to find out a compressed representation of the image frame that is later used by the decoder to get an approximate reconstruction of that image. Hence, acting as somewhat an compression it reduces the dimentionality of the image to the suitable feature vector length for further computations where they are clustered and the images corresponding to the center of the clusters are picked. Since the number of clusters control the number of images in the summary it is on other words $k = \sigma n$. Where k is the number of clusters in the k-means clustering algorithm and sigma is the length regularization hyper parameter. In selection of images, the image corresponding to the feature vector that is nearest to the centroid of the cluster is selected. Since the reconstruction error is supposed to be minimum for sum of pair wise selected euclidean distances between the feature vectors if one of the entries in the pair is fixed to be closest to centroid. To evaluate the optimal number of images in the summary empirical observations can be made regarding the overall sum of pair wise distances in the dataset and then knee can be searched for to optimize the hyper parameter σ for clustering the dataset (Fig. 5).

3.3 Detrimental Point Processes

As point process methods that ensure accurate model of repulsion in the spaces the detrimental point process have found their way into summarization from quantum mechanics and random matrix theory [10]. It is observed that the traditional methods

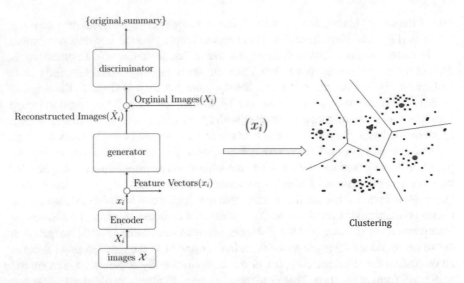

Fig. 5 The above figure describes how deep learning framework can be used for clustering using succinct representation based on feature vectors. Here the summary is evaluated by picking elements closest to centroid of a cluster

such as Markov random fields fail to model accurate description of the affinity in the space in cases where there are negative correlations involved and render themselves computationally intractable. Detrimental Point Process can avoid such pit falls and are key to better and more strategic placements and diversity regulation, therefore catching up attention for the tasks related to summarization [5, 20, 25]. It is quite right fully observed that the probabilistic models are very well now indispensable to the task of machine learning and inference. Therefore, Detrimental Point Process (DPPs) offer an efficient way to infer, sample and marginally understand the feature space in which the data rests and there are very well able to select points from it using custom losses build on DPP. In other words, DPP can be considered as the distribution of the probability of the selection of particular pairs of correlations over the total number of correlations in the space. Where the sub matrix of the correlations is computed using only the elements that shall correspond to the set of the entries of the rows and columns of that sub-matrix. Point process $\mathcal{P}$ serves as an point based probability measure over the subset of the elements of the set $\mathcal{C}$. Now, to reiterate $\mathcal{P}$ as a point-process is the probability measure over all the $2^{\mathcal{C}}$ subsets of $\mathcal{C}$. Then for any random matrix A which can be any subset from the power set of the set $\mathcal{C}$, $\mathcal{P}$ is a DPP if for random subset drawn $\mathcal{K}$ following holds

$$\mathcal{P}(A \subseteq \mathcal{K}) = det(\mathcal{L}_A) \qquad (16)$$

where $\mathcal{L}$ is a $N \times N$ such that $N = |\mathcal{C}|$ symmetric matrix with entries that denote the correspondence and similarities between the two elements of the set using some metric given on which they can be learned. Where in the definition of the $\mathcal{L}_A$ is the set of elements in the matrix $\mathcal{L}$ which are indexed in elements of A in other words $\mathcal{L}_A = [\mathcal{L}_{i,j}] \forall i, j \in A$. Since probabilities are modelled using non negative data it is trivially evident that the matrix $\mathcal{L}$ is positive semi definite matrix. For modelling of the real world data it is important to find the exact probability distributions in several cases and in such cases it becomes important to restrict the case of the system to a particular i.e. of L-ensembles. Therefore the following theorem holds that the point distribution process for the L-ensemble case is proptional to the determinant of the sub matrix.

$$\mathcal{P}_{\mathcal{L}}(\mathcal{K} = A) \propto det(\mathcal{L}_A) \tag{17}$$

The former definition of the point process gave the marginal probabilities of inclusion with respect to the subsets of $\mathcal{K}$ and A. Whereas for the current form of the L-ensemble directly specifies the probability density of the set A for a particular case. Now that we know that the probabilities are proportional to the following value we would like to find the exact probability densities for the distribution.

$$\sum_{A \subseteq \mathcal{K} \subseteq \mathcal{C}} det(\mathcal{L}_{\mathcal{K}}) = det(\mathcal{L} + I_{\hat{A}}) \tag{18}$$

where $\hat{A} = \mathcal{C} - A$ is the complement of subset A and hence $I_{\hat{A}}$ is a matrix of size $N \times N$ with zeros everywhere except for the diagonal elements that belong to $\hat{A}$. Considering the above equation to be true for $A = \mathcal{C}$ we proceed in a form of mathematical induction and obverse

$$det(\mathcal{L} + I_{\hat{A}}) = \sum_{A \cup \{i\} \subseteq \mathcal{K} \subseteq \mathcal{C}} det(\mathcal{L}_{\mathcal{K}}) + \sum_{A \subseteq \mathcal{K} \subseteq \mathcal{C} - \{i\}} det(\mathcal{L}_{\mathcal{K}}) \tag{19}$$

The probability distribution under the point process for the subset A then becomes:

$$\mathcal{P}_{\mathcal{L}}(\mathcal{K} = A) = \frac{det(\mathcal{L}_A)}{det(\mathcal{L} + I)} \tag{20}$$

3.4 Sub-modular Functions

A function $f(.)$ with domain of inputs as sets is said to be submodular for the case where for any element c and the sets $A \subseteq B \subseteq \mathcal{C} - \{c\}$ where $\mathcal{C}$ represents the set of elements corresponding to the ground truth for the corpus. Where folllowing equation holds:

$$f(B \cup \{c\}) - f(B) \leq f(A \cup \{c\}) - f(A) \tag{21}$$

The above equation describes the property of what is called as diminishing returns. In other words, the amount of change in the value of function for addition of a singleton element to the set of existing elements is observed for the case of A where the number of elements is much less than that in B. Therefore aubmodularity has a natural means to converge the ideals of relevance, coverage and diversity imbibed into its formalization. Authors in the paper [7, 21] try to model the task of image corpus summarization using a set of convex submodular functions. To learn the mixture of these submodular functions helps in determination of the corpus summary and its optimal length. It is also proved in the work done by authors in same paper that the cases of clustering and greedy method of selection of the images sorted in increasing order on the basis of their utilities is a special case of this approach. Therefore it is modelled in the form of a submodular scoring function that is experssed as $F_x(A)$ for the set of following submodular functions $f_1, f_2, \ldots f_n$ is a convex sum of same where the vector $x = x_1, x_2, \ldots, x_n$ is convex. In other words it means the $x_i \geq 0 \forall i \in [1, \ldots, n]$ and $\sum_{i=1}^{n} x_i = 1$.

$$F_x(A) = \sum_{i=1}^{n} x_i f_i(A) \tag{22}$$

It is assumed without the loss of the generality that each f_i which is called as a cub-modular component of the original submodular score function is a normalized submodular function. It is supposed to assign measure given the following constraints that $f(\emptyset) = 0$ and $f(C) = 1$. The weights of the learning convex optimization problem are updated in accordance with a designed loss $L(.)$. It can be safely assumed that any summary which is generated with a human level cognition is a righteous summary a given dataset considering all the points have been evaluated by the annotator. Now the competitor summary is assumed to be inferior of the human level summary by a margin which is determined by the loss function in the margin. Thus for the summary $A \in \mathcal{A}$ which represents the human summary and for the summary $A' \in \mathcal{R} - A$ where R is set within $\mathcal{A}$ that satisfies certain budgets for the corpus that is being observed. For example, in many cases the length has to be regularized for the summary given some hyper parameter $c = \sigma|C|$. Also, certain extra budgeting constrains that the summary generated by the system has to follow. R is defined as:

$$R = \{A' \subseteq C \, such that \, |A'| \leq c\} \tag{23}$$

$$F_x(A') + L(A') \leq F_x(A) \, \forall A \in \mathcal{A}, \, A' \in \mathcal{R} - \{A\} \tag{24}$$

where it is assumed that the losses in the system must be in accordance with the standard and be as close to the original human generated summary. Therefore the loss function is restricted between the values of [0, 1]. It is to be noted that the Eq. (24) can be augmented and written in the following form after the normalization.

$$F_x(A) \geq \max_{A' \in \mathcal{R}}[F_x(A') + L(A')] \, \forall A \in \mathcal{A} \tag{25}$$

Which gives us an optimization objective for the minimization of the vector x as following:

$$\min_{x \geq 0, ||x||_1 = 1} \frac{\theta}{2} ||x||_2^2 + \sum_{A \in [\mathcal{A}} \max_{A' \in \mathcal{R}} [F_x(A') + L(A')] - F_x(A)] \qquad (26)$$

4 Quantitative Evaluation Metrics

4.1 Gini Coefficients

As a measure of economic disparity of the country gini coefficient or also known as gini index is a very common metric of measurement of the same. Gini index is a value that ranges between [0, 1] where 1 indicates that there is lot of differences in the selection/ objective selection and 0 indicates vice-versa that there is equal representation from all the sections. Gini-index therefore naturally serves as an evaluation metric for the property of coverage in the system. Therefore it can be concluded that lower the gini index of the summary it ensures more robust and larger coverage in the system. The trends of coverage in an ideally selected system must be constant for whatever the number of images in the system be. Therefore, plotting of the gini index over a range of hyper parameters tells us that how the system responds to the objective of summarization both in generic and task-specific terms. Also it can help in understanding of how its coverage is categorically better than other methods or algorithms for the summarization task. If the gini index for a method is strictly less than other method throughout the change of parameter then the method proves to be categorically employing better coverage ideas as compared to the other algorithm. Gini index is computed for a piece wise differentiable function $F(c)$ for all the nonnegative values of c and with the mean μ of the function gini index is referred as (Fig. 6):

$$G = 1 - \frac{1}{\mu} \int_0^\infty (1 - F(c))^2 dc \qquad (27)$$

4.2 Reconstruction Error

Reconstruction error is used as one of the most prominent measures to find and evaluate the efficacy of the summary. It can be seen the amount of reconstruction error can be controlled by the means of the hyper parameter that allow the fine tuning of the number of outliers in the summary. Higher number of outliers in the summary shall indicate that summary has better coverage and must all better reconstructions of the corpus. In true sense that is the case, as in work done by the authors and reported in their paper Subramanyam et al. [18] they share the findings that the reconstruction error can be controlled by the means of a hyper parameter which defines the number of outliers in the summary by controlling the totally unrelated loss expression from

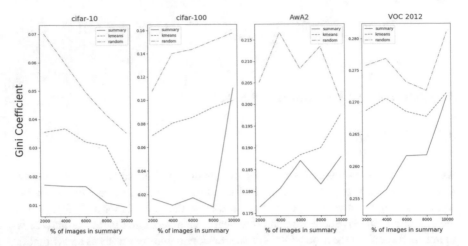

Fig. 6 The above figure describes the value of gini index for different datasets i.e. CIFAR 10, CIFAR 100, AwA2, VOC. Each subfigure compares the gini index for different methods random, clustering and deep learning based models in green, orange and blue respectively. It can be seen for different summary lengths in other words, different σ plotted along x axis. Gini index for the model is consistently lesser than the clustering and randomly selecting summary, as lower gini index shows more equal distribution along classes

the task-specific loss. Given how increasing the parameter increases the number of outliers in the summary and hence sharply decreases the RSME or the reconstruction error of the data set. As the hyper parameter is further increased the effect of task specific loss is now no where to be observed and therefore the reconstruction error rises and becomes constant. While the reconstruction error was high and beta was low it showed that high number of outliers were present in the summary and lot of relevance was jeopardized for the sake of diversity and coverage which resulted in higher values of reconstruction error for small values of the parameter. It can also therefore be noted that a minimum value of reconstruction error is expected when the tuning is done with respect to the change in hyper parameter as the hyper parameter changes the values of task-specific loss. Since for a given task task-specific loss as to ensure maximum coverage it is very likely to include the outliers for that task. In other words the images that set the margin or support the classifier boundary in in hyper dimensional space. Thus the shape of the graph must follow that of the parabola and must be convex in nature which shall help achieve the right number of outliers in the summary. Therefore, at the minimum value of reconstruction error most optimal summary can be observed for any task. To obtain the best value of the hyper parameter a parameter search is performed by running the model for large number of iterations for different values of the parameter. Once the different values of hyper parameter are approached from both the sides i.e. keeping hyper parameter very low and very high we can reach at the mean of the values to search for the parameter. In such cases the optimization can be performed rather quickly and in the

Fig. 7 The above figure shows the change of reconstruction error based on the value of β plotted on the x-axis. It is evident from the profile of plot that for low values of β error is high because lot of outliers are present. Whereas for high values of β is equivalent to only using length regularization

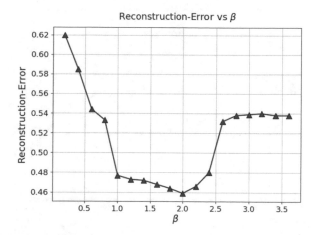

order of $O(log(n))$ iterations as compared to the number of iterations needed for a linear search of the hyper parameter space (Fig. 7).

4.3 Classification or Task Specific Accuracy

When an task specific summarization is done the it has to be done in a manner that the summary is fit as a training corpus for that task. In other words, the summary serves as an small and succinct representation of the dataset which can help train the classifier to a respectable accuracy. The accuracy to which the classifier can be trained or fined tuned using the images from the summary can act as a evaluation measure of how good a summary was in absolute sense. Also speaking in relativistic terms, the ratio of classification accuracy $\phi \in [0, 1]$ of the summary to that of original dataset acts as a robust measure to understand the task specific goodness of the summary. This also serves an important use case as many times it is observed that deep learning based architectures require a huge computational resource and long hours of training to learn the parameters. Fine-tuning such networks which have pretrained parameters and implementing transfer learning on them is also pretty time consuming task. In such cases it becomes really hard to fast prototype and see feasibility of models on datasets. A method to overcome this is to randomly sample the dataset but it is not necessary that the random model will capture all the irregularities in the dataset. Specifically in the case of complex dataset with lot of high resolution images it is tough task to randomly sample images and ensure that all the irregularities and sufficient outliers are captured to test the model. In such cases summarization can help in generation of summaries of fixed length so that the summaries are useful in making the closest possible training of the dataset as compared to training on the full dataset.

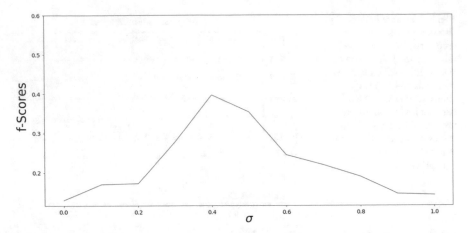

Fig. 8 The above figure is a plot for F-scores with the length of summary for the Diversity 2016 dataset. It shows how efficacy of summary is highest with $0.3 < \sigma < 0.5$ and then later drops because the size of summary increases so recall is supposed to increase but precision drops

4.4 Precision and Recall

Precision and recall is very common means of evaluating the summarization task and other task which serve as a model of inclusion and exclusion [6, 9, 11]. With reference to Fig. 8, it can be observed that the F-scores for the dataset with ground truth annotated and ranked in order we select the number of images in the ground truth to be same as the rank of images. In other words for number of images in summary same number of images are selected from the ground truth and F-scores are calculated. It can be seen that for images with larger values in summary the value of recall increases but overall precision is lost. Hence F-scores reach at maximum for length of summary be 0.3–0.5 times umber of images in corpus.

Even for classification tasks, essentially when the dataset is skewed accuracy can often be misleading. As in case of skewed datasets the algorithm may not learn anything novel and still report accuracy greater than 90% since it could be the case that the detection for the cases is very rare and the dataset may not have enough examples of positives. For example, in case of cancer prediction problem it could be very well the case that 99.9% of the people may not have cancer. So even if the algorithm doesn't learn to detect cancer and only labels no cancer found for every entry it would do well on the 99.9% of the dataset and have 99.9% accuracy. Therefore, a confusion matrix is constructed to find our different possibilities and precision and recall for the same can be reported. In the case of summarization with respect to image corpus precision recall and F-scores are calculated as follows:

$$Precision = \frac{\text{Number of images in summary and ground truth}}{\text{Number of images in summary}} \quad (28)$$

$$Recall = \frac{\text{Number of images in summary and ground truth}}{\text{Number of images in Ground Truth}} \qquad (29)$$

$$\frac{1}{Fscore} = \frac{1}{2} \left(\frac{1}{Precision} + \frac{1}{Recall} \right) \qquad (30)$$

4.5 V-Rogue

V-Rogue is an metric that is developed on the lines of Rogue metric which is common way of assessment of the text summarization algorithms and their performance. An analogous method to evaluate the performance of image collection summarization methods was coined a V-Rogue. Which is as follows:

$$r_S(X) = \frac{\sum_{y \in \mathcal{Y}} \sum_{S \in S} \min(c_y(X), C_y(S))}{\sum_{y \in \mathcal{Y}} \sum_{S \in S} C_y(S)} \qquad (31)$$

where S is the human level generation of the summaries and y is the set of features that are desired in the summary and $C_y(X)$ is the count of the occurrence of those features in the summary X which is generated by the competitor.

5 Qualitative Evaluation Metrics

5.1 T-Sne Visualization

For observing datasets with diverse set of images and lot of different features T-Sne provides a robust mechanism to find our the underlying relationships with the different set of images. T-Sne maps the similar images in to a similar cluster while at the same time repelling dissimilar images and therefor helping underline the commonalities in the data points in a dimension space that is very small as compared to the rest of the space. Tsne brings it differently from other dimensionality reduction methods such as eigen value decomposition, PCA or SVD. It is because it not only does find a representation where similar images are brought together in distance of metric chosen but the dissimilar classes are also repulsed from each other. It makes it more discriminative and helps by working in adversarial manner (Fig. 9).

5.2 Visualization by Saliency Within an Image

Visualization of the efficacy of model can also be done by making summarization work as an attention model or a saliency detection model. A image can be broken into

Fig. 9 Karaparthy style tsne
plot of one percent of images
selected by the model form
the AwA2 dataset. Each
image is embedded at the
point within subspace where
it is located with respect to
rest of images. Please zoom
in for better visualization

multiple blocks and patches of a pre-defined size. Now each block can be up-sampled
to the given dimensions for the input of the summarization model. Now when each
patch or block of image is treated as an equivalent of an image in corpus. Then the
summarization model runs for a fixed hyper parameter and outputs fixed number
of images in the summary. Now for up scaled images corresponding to a block the
block is stitched on the white image block at the location from where it was cropped.
It gives an overall idea of the kind of blocks which the algorithm gives importance.
In other words, it describes the blocks in summary to be the patches that are integral
for understanding of the image.

5.3 *Visualization by Ranking with Ground Truth*

Visualization of the efficacy of model can also be done in qualitative manner when we
have ground truth defined for the dataset. The images when ranked in terms of their
scores, i.e certain number of images picked with highest scores can be compared with
the ranked images in ground truth based on their relevance. This helps us visualize
not just only in the terms of intersection which the images selected by the model have
in common to the ground truth but also the degree to which the scoring mechanism
agrees with ground truth ranking. Which in turn helps gain deeper insight into efficacy
of the model in the terms of relevance. With reference to Figs. 10, 11, 12 and 13 it can
be observed for two different feature vectors in different deep learning models can
be compared with each other and also with the ground truth. For two chosen classes
from the set different classes consisting of many events and objects in the Diversity
2016 qualitative evaluation is presented.

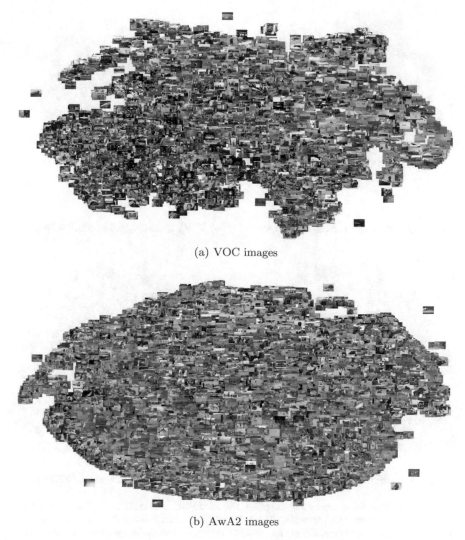

(a) VOC images

(b) AwA2 images

Fig. 10 The above figure talks about the t-SNE plots for VOC and AwA2 dataset. **a** is t-SNE for full voc dataset and **b** is t-sne for full AwA2 dataset. Images are embedded at the corresponding locations. Please zoom in for better visualization

6 Dataset Design for Summarization

The design of dataset is one of the most challenging and subjective areas. This is somewhat the reason because of which there aren't any enough datasets which have been made public for the task of image collection summarization. Most of the research usually focuses on the collection of datasets from the corpus formed using images scraped from web in other words Google and Flickr are common places from

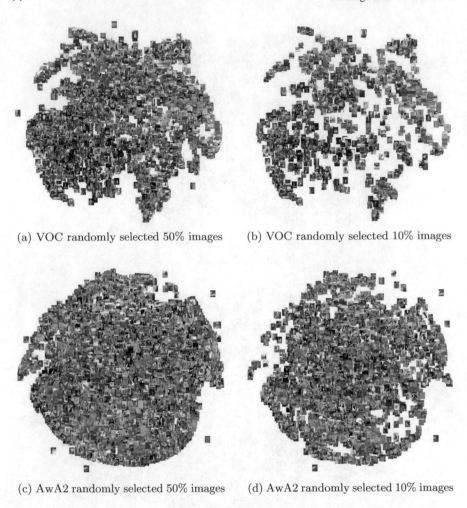

(a) VOC randomly selected 50% images   (b) VOC randomly selected 10% images

(c) AwA2 randomly selected 50% images   (d) AwA2 randomly selected 10% images

Fig. 11 The above figure talks about the t-SNE plots for VOC and AwA2 data set **a** is a randomly generated summary for VOC dataset at 50%. **b** is randomly generated summary VOC dataset with 10% of images. Similarly **c** and **d** are for AwA2 dataset respectively. Images are embedded at the corresponding locations. Please zoom in for better visualization

which images are relevantly scraped to form the image collection summarization datasets. Then the data is subjected to a ground truth of binary classification using certain group of users who validate the corpus. Sometimes the images are subjected to verification from amazon mechanical Turks where Turks are instructed to work in a fashion to label the images as best fit for the summary. The problem with the statement is that the task is not that objective for the human also as compared to the tasks of person-reidentification, image classification and object detection. The advantages of the method to design a dataset for the task in this manner is that the current design of dataset supports us to leverage the ground truth in the dataset for

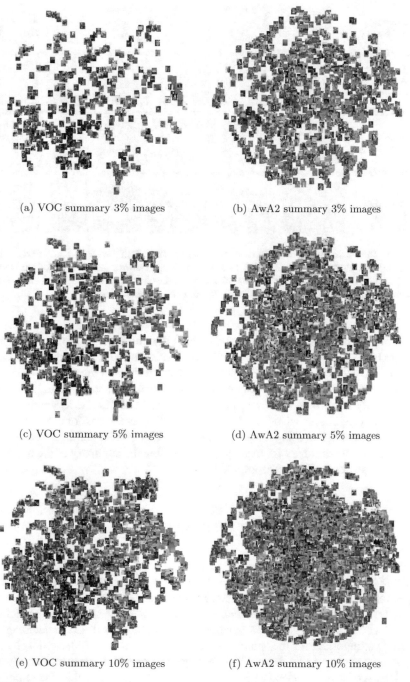

(a) VOC summary 3% images

(b) AwA2 summary 3% images

(c) VOC summary 5% images

(d) AwA2 summary 5% images

(e) VOC summary 10% images

(f) AwA2 summary 10% images

Fig. 12 The above figure talks about the t-SNE plots for VOC data set **a** is summary for VOC dataset at 3%. **c** and **e** is the summary for VOC dataset, 5% and 10% of images. **d** and **f** is t-SNE for AwA2 dataset. Images are embedded at the corresponding locations. Please zoom in for better visualization

Fig. 13 The above figure shows Top 5 images selected from both covnet and auto-encoder based deep learning model features. The first two show a comparison for clouds class of the Diversity 2016 data set and the last two rows show images from the class of event Tour De France. The top 5 images can also be considered as an summary for event summarization

multiple evaluation metrics which shall tell us the goodness of the summary. Such design has certain applications in recommendations to the users who have made or curated the dataset as a common metric can be learned about the choices or interests of the users which can then be projected to a another feature space. Alternatively, the different domain preference for the targeted user can be referred as a joint distribution of the two preferences where the distribution of his preferences for the original case is already established using the ground truth. In other words, helping to recommend set of images as summary for a query when knowing the summary of the user which was provided in the dataset as ground truth.

7 Discussion and Further Challenges

It is important for the summarization research to solve multiple existing challenges that the summarization of corpus poses. Various tools and standards need to be developed for the summarization to be dealt as a problem in a manner that helps efforts be more reproducible in different scenarios. Authors in some of their published work describe how the intent for summarization holds an important role in deciding what should be important for the summary. Also, certain new set of ideas must be more rigorously explored like detrimental point processes which provide a new way to model diversity in the selection procedure. It also becomes to realize the importance of more robustness into the set of features that we are working on, more specifically it is important to have more deep learning based models that try to solve the problem of image corpus summarization. While introduction of deep learning based approach to image corpus summarization it results in an very important application of summarization. In other words, with the current state of the art models in deep learning have close to intractable number of parameters that need to be learned. It is known that even for fine-tuning such models lot of compute and data is required and sometimes, the models have to run for months to be learned that they perform sufficiently well on the test data. Also, it is well known that the deep learning based solutions require an enormous amount of data in lieu of its training. Therefore, summarization of the data sets in such manner that they retain the properties well enough in their summaries to train a classifier to reasonably comparable accuracy as compared to the state of the art. This will give deep learning pipeline an acceleration and remove the barrier of compute and access to massive datasets. As faster prototyping for models can be done also when time and compute are an issue working on smaller datasets shall provide an edge and make application of deep learning more and more ubiquitous.

References

1. Aner A, Kender JR (2002) Video summaries through mosaic-based shot and scene clustering. In: European conference on computer vision. Springer, pp. 388–402 (2002)
2. De Avila SEF, Lopes APB, da Luz Jr A, de Albuquerque Araújo A (2011) Vsumm: a mechanism designed to produce static video summaries and a novel evaluation method. Pattern Recogn Lett **32**(1), 56–68 (2011)
3. Denton T, Demirci MF, Abrahamson J, Shokoufandeh A, Dickinson S (2004) Selecting canonical views for view-based 3-d object recognition. In: Proceedings of the 17th International Conference on Pattern Recognition, 2004. ICPR 2004. vol. 2. IEEE, pp. 273–276 (2004)
4. Freeman WT (1994) The generic viewpoint assumption in a framework for visual perception. Nature 368(6471):542
5. Gong B, Chao WL, Grauman K, Sha F (2014) Diverse sequential subset selection for supervised video summarization. In: Advances in neural information processing systems, pp. 2069–2077
6. Gygli M, Grabner H, Riemenschneider H, van Gool L (2014) Creating summaries from user videos. In: European conference on computer vision. Springer, pp 505–520
7. Gygli M, Grabner H, van Gool L (2015) Video summarization by learning submodular mixtures of objectives. In: Proceedings of the IEEE conference on computer vision and pattern recognition, pp. 3090–3098
8. Hall PM, Owen M (2005) Simple canonical views. In: BMVC, pp 839–848
9. Khosla A, Hamid R, Lin CJ, Sundaresan N (2013) Large-scale video summarization using web-image priors. In: Proceedings of the IEEE conference on computer vision and pattern recognition, pp 2698–2705
10. Kulesza A, Taskar B et al (2012) Determinantal point processes for machine learning. Found Ttrends® Mach Learn 5(2–3):123–286
11. Li Y, Merialdo B (2010) Multi-video summarization based on video-mmr. In: 11th International workshop on image analysis for multimedia interactive services WIAMIS 10. IEEE, pp 1–4
12. Mahasseni B, Lam M, Todorovic S (2017) Unsupervised video summarization with adversarial lstm networks. In: Proceedings of the IEEE conference on computer vision and pattern recognition, pp 202–211
13. Palmer S (1981) Canonical perspective and the perception of objects. Atten Perform 135–151
14. Plummer BA, Brown M, Lazebnik S (2017) Enhancing video summarization via vision-language embedding. In: Proceedings of the IEEE conference on computer vision and pattern recognition, pp 5781–5789

15. Savakis AE, Etz SP, Loui, AC (2000) Evaluation of image appeal in consumer photography. In: Human vision and electronic imaging V, vol 3959. International Society for Optics and Photonics, pp. 111–121
16. Simon I, Snavely N, Seitz SM (2007) Scene summarization for online image collections. In: IEEE 11th international conference on computer vision, 2007. ICCV 2007. IEEE, pp 1–8
17. Singh A, Sharma DK. Dlk-sum: image collection summarization using deep learning based architecture and k-means clustering. https://anurag14.github.io/
18. Singh A, Virmani L, Subramanyam A Summary generation for image corpus using generative adversarial networks. https://anurag14.github.io/
19. Sinha P, Mehrotra S, Jain R (2011) Summarization of personal photologs using multidimensional content and context. In: Proceedings of the 1st ACM international conference on multimedia retrieval. ACM, p. 4
20. Szegedy C, Liu W, Jia Y, Sermanet P, Reed S, Anguelov D, Erhan D, Vanhoucke V, Rabinovich A (2015) Going deeper with convolutions. In: Proceedings of the IEEE conference on computer vision and pattern recognition, pp 1–9
21. Tschiatschek S, Iyer RK, Wei H, Bilmes JA (2014) Learning mixtures of submodular functions for image collection summarization. In: Advances in neural information processing systems, pp. 1413–1421
22. Weinshall D, Werman M, Gdalyahu Y (1994) Canonical views, or the stability and likelihood of images of 3d objects. In: Image understanding workshop, pp. 967–971
23. Yang C, Shen J, Peng J, Fan J (2013) Image collection summarization via dictionary learning for sparse representation. Pattern Recogn 46(3):948–961
24. Zhang K, Chao WL, Sha F, Grauman K (2016) Summary transfer: exemplar-based subset selection for video summarization. In: Proceedings of the IEEE conference on computer vision and pattern recognition, pp. 1059–1067
25. Zhang K, Chao WL, Sha F, Grauman K (2016) Video summarization with long short-term memory. In: European conference on computer vision. Springer, pp. 766–782
26. Zhao J, Mathieu M, LeCun Y (2016) Energy-based generative adversarial network. arXiv:1609.03126

Deep Learning Based Semantic Segmentation Applied to Satellite Image

Manami Barthakur and Kandarpa Kumar Sarma

Abstract Satellite images carry essential information required for range of applications. Information extraction from satellite images is a challenging issue and requires a host of support. Further automation of information extraction, reliability and decision making regarding content are essential and vital elements. Of late, among the learning based approaches deep neural network (DNN) supported methods have been to efficient and reliable. In this work a DNN based approach optimized for satellite images have been discussed. The chapter includes the details of the techniques adopted, including the DNN topologies, experiments performed, results and related discussion. The work includes details of a specially configured and trained CNN topology which is found to be suitable for satellite image segmentation. Finally an approach based on deep learning for semantic segmentation in satellite images is also proposed. From the experimental results it was seen that the method is appropriate for real world and reliable.

Keywords Semantic segmentation · CNN · Deep learning · Satellite image · SegNet · Image segmentation

1 Introduction

A versatile tool for exploring the Earth is remote sensing. Satellite images, also known as remotely sensed images, are the data recorded by sensors from a very small portion of the Earth's surface. In these images, to capture the spatial and spectral relations of objects and materials perceptible at a distance, different instruments or sensors are used. Satellite images are mainly used in geographic information systems (GIS) [1]. The GIS systems which are used for classification are useful for cartography. The

M. Barthakur (✉) · K. K. Sarma
Department of Electronics and Communication Engineering, Gauhati University, Guwahati, Assam, India
e-mail: manamibarthakur@gmail.com

K. K. Sarma
e-mail: kandarpaks@gmail.com

© Springer Nature Switzerland AG 2020
J. Hemanth et al. (eds.), *Data Visualization and Knowledge Engineering*,
Lecture Notes on Data Engineering and Communications Technologies 32,
https://doi.org/10.1007/978-3-030-25797-2_4

intensity of the pixels in satellite images with low resolution, is enough to individually classify each of them. On the other hand, image classification for high resolution images is more difficult, since the level of detail and the heterogeneity of the scenes is raised. The high resolution images contain much information inside it. A lot of characteristics which are associated to nature may be there in a satellite image such as color, shape, texture or structure, density [2]. The existing Geographical Information Systems (GIS) classification systems use the basic classification methods for low and high resolution images. These systems give satisfictory results for low resolution images, but with new high resolution images, these same basic methods cannot provide satisfactory results. Now a days, to improve classification accuracy, scientists and researchers have made much efforts towards the development of advanced classification approaches and techniques. To classify different regions in satellite images, semantic segmentation is frequently used. Here each pixel of an image is associated with a class of what the object is being represented [3]. It is essential for many image analysis tasks. The semantic segmentation differentiates between the objects of interest and their background or other objects. Semantic segmentation is also used in many applications such as autonomous driving, industrial inspection, medical imaging analysis, military reconnaissance, weather forecast, land use patterns, crop census, ocean resources and ground water studies etc. [3].

An ideal algorithm of image segmentation is expected to segment objects which are unknown or new. Several approaches such as Semantic Texton Forest [4] and Random Forest based classifiers [5] for semantic segmentation are exits in the literature. Many of these approaches depend on the characteristics of images which can be measured. For this reason, these methods work well in some of the cases and do not work well in others. Again, there are some unintentional alterations in the images due to noise, image intensity non-uniformity, missing or occluded portion in the image etc. Therefore, for segmentation in complex images, methods based on prior knowledge may be more suitable then other approaches. For these reasons neuro-computing methods with learning algorithm have been applied a lot in the literature [6–8].

Presently, Deep learning is becoming popular since it is very useful for real-world applications due to the efficiency and reliability it generates. It working is based on in depth learning of features and mapping these to probable output state. Most of the Semantic segmentation problems are performed using deep networks, such as Convolutional Neural Networks (CNNs) [7–9]. In terms of efficiency and accuracy these methods are surpassing other methods extensively.

Several learning based methods have been developed for semantic segmentation until now. Based on the object detection results Regions with CNN [10] (RCNN) performs the semantic segmentation. It computes CNN features by utilizing selective search to take out extensive amount of object proposals. Uijlings et al. [11] reported for bilinear up-sampling to pixel-dense outputs, the deconvolutional layers which follow the multilayer outputs to have combined coarse, information of high layer with fine and information of low layer. In [12], a deep learning based vehicle detection method in satellite images is proposed.

In this chapter, some of the recent methods developed using deep learning based approaches for semantic segmentation in satellite images are discussed. It includes the details of adopted for deep learning topologies, experiments performed, results and related discussion. The works includes a specially configured and trained CNN topology which is found to be suitable for satellite image segmentation. Finally an approach based on deep learning for semantic segmentation in satellite images is also proposed. The method comprises of training and formation of a SegNet where the input images are satellite images. In the work, the target to the network was taken from the output of algorithm of K-means clustering with their label of the required region of interest (ROI). The ROI is reinforced by the learning of the CNN which is later used for extraction and identification. Then, with the output of the SegNet as input and the pixel values of various ROI as target a neuro-computing structure is trained as classifier to segment the various ROIs. Here, the neuro-computing structure is a Multi Layer Perceptron (MLP). The MLP is trained with (error) Back Propagation learning.

The proposed method uses raw images as input and has no dependence on hand crafted features. Further, the approach does not uses any pre-processing techniques such as region growing, region split-merge etc. The method automates the approach of information extraction from satellite images. Raw image inputs are used to configure and train the SegNet, which is a deep convolutional network with Encoder-Decoder architecture. It is trained with (error) Back Propagation learning. The output of the SegNet is used for classification and decision using the multi layered neuro-computing classifier, the MLP. From experimental results it can be concluded the method is appropriate for real world and reliable.

In this Chapter, Sect. 2 includes, the review on some of the recent methods developed using deep learning based approaches for semantic segmentation. The proposed algorithm is presented in Sect. 3. Then Sect. 4 includes the experimental results and at last, the conclusion of the work is discussed in Sect. 5.

2   Review of Recent Methods on Deep Learning Based Semantic Segmentation

Several algorithm have been developed for semantic segmentation based on deep learning. Some of the recent methods are discussed here.

Long et al. [13] developed a "fully convolutional" network which with efficient inference and learning can produce correspondingly-sized output by taking input of arbitrary size. It was later applied to spatially dense prediction tasks, deep classification networks (GoogLeNet [14], VGG net [15], and AlexNet [16]) and converted them into fully convolutional networks. Using fine tuning [17] they transferred learned representations to the segmentation task.

Badrinarayanan et al. [6] presented a deep fully convolutional neural network. The network developed for semantic pixel-wise segmentation which has been termed SegNet. It has an encoder network and a corresponding decoder network. The encoder and decoder is followed by a pixelwise classification layer.

Ronneberger et al. [18] developed a method based on deep learning for biomedical image segmentation which has been termed as U-Net. It has been modified and extended to the fully convolutional network [13] architecture such that the network works with very less number of training images to give more precise segmentations.

Yoshihara et al. [19] proposed a semantic segmentation method for satellite images using fully convolutional network. The architecture of the network comprises of an encoder network followed by a corresponding decoder network like [18, 20]. The input size to the network has been changed from the employed value in [20] to 256 × 256. The encoder network architecture was same with of a convolutional network.

Patravali et al. in [21] developed a 2D and 3D segmentation method for fully automated segmentation of cardiac MR image. The method is based on Deep Convolutional Neural Networks (CNN). They had developed the 2D segmentation model architecture as in [18].

Langkvist et al. [22] introduced an CNN based approach for per-pixel classification of satellite.

Jegou et al. in [20] extended the DenseNet architecture [23] for semantic segmentation. They extended the network to fully convolutional networks (FC-DenseNets), while reducing the feature map explosion.

Chaurasia et al. [24] proposed a deep neural network (LinkNet) which allows it to learn without any significant increase in number of parameters. Their proposed deep neural network architecture attempted to efficiently share the information learnt by the encoder with the decoder after each downsampling block.

In [25] He et al. developed the Mask R-CNN architecture which has been an extension of popular Faster R-CNN architecture. The developed the architecture by changing required parameters to perform semantic segmentation.

3 The Proposed Method

The system model for the proposed work is shown in the Fig. 1. Each block of the work is described in the following sections. The input to the system is outlined in Sect. 3.1. Then the k means clustering algorithm (KMCA) is described in Sect. 3.2.

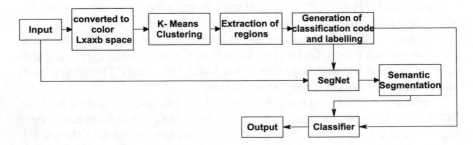

Fig. 1 Block diagram explaining the proposed work

In Sect. 3.3 a brief introduction to the SegNet network is given. Selection of RoI is discussed in Sect. 3.4 and finally in Sect. 3.5 the neuro computing classifier which is trained using the output of SegNet as input and the RoI as target is explained.

3.1 Input and Dataset

The input in this work are Satellite images. The images are taken from United States Geological Survey [26], Inria database [27] and deep globe database [28], which are high resolution aerial images. The Geometric resolutions of the images are 0.25, 0.3 and 0.4 m. The capability of the satellite sensor to capture a portion of the surface of earth is called Geometric resolution and is expressed as Ground sample distance. Multiple sets of 100 multi colored images of varying sizes are taken for the work during training, Similarly for testing another set of 100 images are taken. For training the classifiers 100 images of each classes for multiple configuration are taken and average results reported.

At first, 50% of the original image is taken as input by resizing it. Then it is converted to L × a × b Color Space. The color space has layers of a luminosity 'L', chromaticity 'a' and chromaticity 'b'. The color information can be obtained from 'a*' and 'b*' layers. Therefore, for further processing, the 'a*' and 'b*' values of the converted image pixels are used.

3.2 K-Means Clustering Algorithm

The KMCA is applied to group different similar regions of the image. To group the data so that the similar objects fall on one cluster and dissimilar on other, KMCA is used. It is an algorithm which classifies or groups the objects in image into K different groups. K is a positive number. By minimizing the Euclidean distances between data and its cluster centroid, the grouping is done. Let us consider the image that is to be cluster is of resolution of x × y. Let the input pixel to be cluster be p(x, y) and the cluster centers be c_k. The KCMA is following as [29]:

Step 1. The cluster number k, and their centre be initialized.
Step 2. The Euclidean distance d, between the initialised center and each image pixel be calculated as

$$d = \|p(x, y) - c_k\| \qquad (1)$$

Step 3. Based on distance d, to the nearest centres, all pixels in the image be assigned.
Step 4. New position of the centre be recalculated after assigning all pixels using the relation given below.

$$c_k = \frac{1}{k} \sum_{y \in c_k} \sum_{x \in c_k} p(x, y) \tag{2}$$

Step 5. The process be repeated until the it achieved minimum error value.
Step 6. Then the pixels in the cluster be reshaped into image.

The index returned by the algorithm corresponds to the clusters in the image. Using these index values labeling of every pixel in the image is done. Then the ROI of the image are extracted, according to the labeled pixels.

3.3 SegNet

It is a deep fully convolutional neural network [6]. The architecture developed for semantic pixel-wise segmentation. It has an encoder network. The encoder network is followed by a corresponding decoder network. Then a final pixelwise classification layer follows the network (Fig. 2).

The layers in the encoder and decoder network are explained below:

Convolutional layer. Convolutional layer is a bank of simple frs. It has learnable parameters. Let us consider the size of the input layer X be $W_1 \times H_1 \times D_1$. This layer convolves with the filter bank. Here the stride be S and padding P units. Then result of the operation will be an output Y whose size will be $W_2 \times H_2 \times D_2$ as shown in Fig. 3. The output at spatial position (i, j) of Y formulates as:

$$Y_{ij} = w \times N_{ij} + b \tag{3}$$

Fig. 2 SegNet Architecture [6]

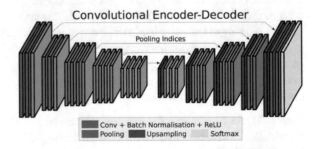

Fig. 3 Illustration of convolutional layer: Convolution performed using input of size $W_1 \times H_1 = 5 \times 5$, with filter size (receptive field, F) 3, stride S = 2 and padding P = 0. The output Y will be of size 3×3 [30]

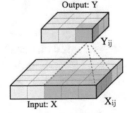

where b and w are the bias and weights respectively which are learnable parameters of the layer, N_{ij} is the corresponding receptive field. The dimension of the spatial output of the layer are given by $W_2 = (W_1 - F + 2P)/S + 1, H_2 = (H_1 - F + 2P)/S + 1$. Here F is the size of the receptive field, or spatial size of the filters. In this work, the convolution filters are of size 3×3.

The output volume neuron size is $F \times F \times D_1$. By the method of parameter sharing the learnable weights and bias of Y are shared across spatial location of X.

Non-Linear Function Layer. A non-linear function layer is situated after the convolution layer. This layer is called an activation function. The Activation function introduces non-linearity in the network. The rectified linear unit (ReLU) function [30], $f(x) = \max(0, x)$, is the activation function which is most commonly used.

Pooling Layer. Using the MAX operation, the Pooling Layer resizes every depth slice of the input spatially and independently as shown in Fig. 4. The filters of size 2×2 with a stride of 2 is the most common form of a pooling layer. Along both width and height, the pooling layer discard 75% of ReLU layer output by downsampling by 2. The depth dimension will not changed. The pooling layer output will be $W_2 = (W_1 - F)/S + 1, H_2 = (H_1 - F)/S + 1, D_2 = D_1$. Here the input volume is of size $W_1 \times H_1 \times D_1$. F is filter size and S is the Stride.

Unpooling layer. The locations of the maxima in the output of each pooling area is recorded in the unpooling operation. In the deconvolution, by preserving the structure of the stimulus, the unpooling operation places the values from the layer of pooling into appropriate locations, using these recorded location.

Deconvolution layer. The deconvolution layer, also known as the transposed layer, is shown in Fig. 4. For upsampling operation is layer is commonly used. The input to the layer is first up-sampled with stride S and padding PThen, with a filter bank, convolution is performed in the input to up-sampled it with receptive field of size F. It is inverse of convolution. The parameters of the filter can be set to be learned.

Softmax classifier. Softmax layer also referred to as a normalized exponential function. It squashes a k-dimensional vector(z) to a k-dimensional vector $(\sigma(z))$ in the range $(0, 1)$ that add up to 1. The equation is as follows

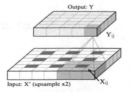

Fig. 4 Illustration of deconvolutional layer: deconvolution performed using input of size $W_1 \times H_1$ = 3×3, with filter size (receptive field, F) 3, stride $S = 2$ and padding $P = 1$. The output Y will be of size 5×5 [30]

$$\sigma : \mathbb{R}^k \rightarrow (0, 1)^k$$

$$\sigma(z)_j = \frac{e^{z_j}}{\sum_{k=1}^{k} e^{z_k}} \qquad (4)$$

The Softmax classifier uses the cross-entropy loss. Each pixel is examined by this loss individually, comparing the class predictions to the target.

In SegNet, with a filter blank each encoder performs convolution. This produces a set of feature maps. The feature maps are then batch normalized. An element-wise ReLU, (max (0, x)) is performed after batch normalization. Following that, max-pooling with a non-overlapping window with stride 2 is performed. The window is of size 2×2. Then the output of max-pooling is sub-sampled by 2. The max pooling indices are stored and used in the decoder network. Using the stored max-pooling indices, the decoder network upsample the input feature maps. This step produces sparse feature maps. To produce dense feature maps the feature maps are convolved with a trainable decoder filter bank. Finally, trainable soft-max classifier is used to process the high dimensional feature representation of the final decoder output. Each pixel is classified by this soft-max classifier independently. The resultant segmentation is the class with maximum probability at each pixel [6].

Using the ROI as input and the label image of each ROI as target, the SegNet is trained with Stochastic Gradient Decent (SGD) using backpropagation as a gradient computing technique. This method is adopted here since the weight update technique has the ability to train end to jointly optimize the network weights.

The algorithm for training the SegNet is given below. In algorithm1 and algorithm2 training of encoder and decoder are described and in algorithm3 batch normalization is presented [31].

Algorithm 1. Training an encoder: Training start with a input image of size $X_1 \times H_1 \times D_1$. The size of output is $X_2 \times H_2 \times D_2$. Here, the cost function for mini-batch is denoted by C. The value λ denotes the learning rate decay factor and L denotes the numbers of layers. the batch-normalize activations are denoted by $BatchNorm()$. The $ReLU()$ performs activation (max(0, x)) of the given input. The $Conv()$ performs operations of convolution with a filter size $= 3, S = 1$(stride) and $P = 1$(zero padding). The pooling operation is performed by The $Pool()$ with Max pooling filter with kernel size $= 2, S = 2$(stride) and $P = 0$(zero padding).

Require: A mini-batch of inputs and targets (p, t), previous weights W, previous Batch Normalization parameters θ, weights initialization coefficients from γ, and previous learning rate η.

Ensure: weights updating W^{t+1}, Batch-Normalization parameters updating θ^{t+1} and learning_rate updating η^{t+1}.

Step1: Input $X_1 \times H_1 \times D_1$ size image
Step2: **for** $k = 0$ to $L - 1$

Repeat:
   1: $W_{bk} \leftarrow BN(W_k)$
   2: $W_{rk} \leftarrow ReLU(W_k)$

3: $W_{ck} \leftarrow Conv. (W_{rk}, W_{bk})$
4: $Pl_{kmask}, Pl_k \leftarrow Pool(W_{ck})$
5: **if** $k \le L$ **then**
$\quad W_k \leftarrow Pl_k$
return Pl_{kmask}

Step3: Stop

Algorithm 2. Decoder training: Input pooling mask size is of $X_2 \times \leftarrow H_2 \times \leftarrow D_2$. The output is of size $X_D \times \leftarrow H_D \times \leftarrow D_D$. Here, the cost function for mini-batch is denoted by C. The value λ denotes the learning rate decay factor and L denotes the numbers of layers. The function that up-samples the given inputs is denoted by *Upsample*(). As the encoder configurations, The *ReLU*() and *Conv*() will perform the same operations. To output the class probabilities *SoftMax*() is used which is a multi-class softmax classifier.

Require: A mini-batch of feature maps extracted from encoder network of size $X_2 \times H_2 \times D_2$. Previous weights W, previous Batch Normalization parameters θ, weights initialization coefficients from γ, and previous learning rate η.

Ensure: weights updating W^{t+1}, Batch-Normalization parameters updating θ^{t+1} and learning rate updating η^{t+1}.

Step1: Input Pl_{kmask}, Pl_k
Step2: **for** $k = L$ to 1

Repeat:
1. $F_{dk} \leftarrow Upsample(Pl_{kmask}, Pl_k)$
2. $F_{dbk} \leftarrow BatchNorm(F_{dk})$
3. $F_{dek} \leftarrow ReLU(F_{dk})$
4. $F_{dck} \leftarrow Conv(F_{dek}, F_{dbk})$
5. **if** $k > 1$ **then**
$\quad Pl_{k-1} \leftarrow Fdck$
else return
$\quad SoftMax(F_{dck})$

Step3: Stop

Algorithm 3. Batch normalization [31]: Let us consider the normalized values and their corresponding linear transformations be $\hat{x}_{1...m}$ and $y_{1...m}$ respectively. $BN_{\gamma\beta}$: $x_{1...m} \rightarrow y_{1...m}$ refers the transformation. For numerical stability, the value $\in$ is added, which is a constant, to the mini-batch variance.

Require: Values of x over a mini-batch: $B = x_{1...m}$; Parameters to be learned: γ, β

Ensure: $y_i = BN(x_i, \gamma, \beta)$

$$\mu_B \leftarrow \frac{1}{m} \sum_{i=1}^{m} x_i \quad //\text{mini-batch mean}$$

$$\sigma_B^2 \leftarrow \frac{1}{m} \sum_{i=1}^{m} (x_i - \mu_B)^2 \quad //\text{mini-batch variance}$$

$$\hat{x}_i \leftarrow \frac{x_i - \mu_B}{\sqrt{\sigma_B^2 + \epsilon}} \quad //\text{normalize}$$

$$yi \leftarrow \gamma \hat{x}_i + \beta \equiv BN(x_i, \gamma, \beta) \quad //\text{scale and shift}$$

3.4 Selection of ROI

The output of the SegNet will be a image where each pixel of the image is associated
with a class of what is being represented. From the output of the SegNet the required
ROI is selected. The pixel values of the different required regions are taken as the
ROI. To segment the sea region pixel values of that region is kept as it is and other
values are kept as 255. Thus the pixel values of a particular region will be target for
the classifier to segment out the region.

3.5 Feed Forward Neuro Classifier

With the output of the SegNet as input and the pixel values of various ROI as target
a neuro-computing structure is trained to segmentment the various ROI. Here the
neuro-computing structure is a Multi Layer Perceptron (MLP). The MLP is trained
with (error) Back Propagation (BP) algorithm and used as classifier at the last end.

To learn applied patterns, the classifier is configured. The process through which
the classifier learns is called training. The classifier is trained with BP algorithm.
Depending on the BP algorithm, the weights between the layers of the classifier are
updated. Till the performance results meet its goal, this adaptive updation of the
classifier is continued. The steps are as below [32]:

- Initialization: With random values between [0, 1] weight matrix W is initialized.
- The training samples: $I_p = [i_{p1}, i_{p2}, \ldots i_{pn}]$ is the input and $T_p = [t_{p1}, t_{p2}, \ldots t_{pn}]$
  is the desired output. The hidden nodes are calculated as,

$$N_{py}^q = \sum_{x=1}^{n} w_{yx}^q I^{px} + f_y^q \tag{5}$$

The hidden layer output of the classifier are computed as-

$$O_{py}^q = f_y^q(N_{py}^q) \tag{6}$$

where, f(X) is a function which depends on activation function.

The values of the output nodes of the classifier are computed as-

$$O_{pl}^r = f_l^r(N_{py}^r) \tag{7}$$

- Computation of errors: The errors between the output nodes of the classifier and target are calculated as

$$E_{ye} = T_{ye} - O_{ye} \tag{8}$$

The MSE i.e. the mean square error is computed as-

$$MSE = \frac{\sum_{y=1}^{N} \sum_{e=1}^{n} E_{ye}^{2}}{2N} \tag{9}$$

Then output layer error is computed as follows-

$$\partial_{pm}^{r} = O_{pm}^{r}(1 - O_{pm}^{r})E_{pe} \tag{10}$$

And the hidden layer error is computed as-

$$\partial_{pm}^{q} = O_{pm}^{q}(1 - O_{pm}^{q}) \sum_{y} \partial_{py}^{r} \omega_{ym} \tag{11}$$

- Updating of Weights: The weights connecting the hidden layer and output layer are updated as

$$\omega_{my}^{r}(t+1) = \omega_{my}^{r}(t) + \eta \partial_{pm}^{r} O_{py} \tag{12}$$

where η is called the learning rate of the classifier ($0 < \eta < 1$).

Again the weights connecting the input layer and the hidden layer are updated as

$$\omega_{yx}^{q}(t+1) = \omega_{yx}^{q}(t) + \eta \partial_{py}^{q} I_{x} \tag{13}$$

One training cycle forms one epoch. Till the error meets the performance goal, the process of updating of weights is repeated.

The output of the classifier will be the values of pixel of the required area that is to be segmented.

The process map for the work is shown in Fig. 5. Here the feature map obtained as SegNet output is used to train a feed forward classifier with the labeled ROI as target.

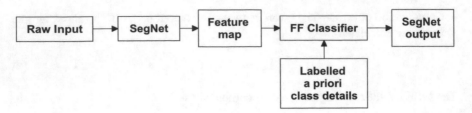

Fig. 5 The process map for the proposed work

4 Experimental Results and Discussion

The experimental results are discussed in the following sections. In Sect. 4.1 the results obtained for the SegNet network are discussed. The results obtained from the classifier are discussed in Sect. 4.2.

4.1 Results Obtained from SegNet

The method is tested using two images—"Satellite image 1" and "Satellite image 2". The two images used for testing are shown in Figs. 6 and 10. These images are used to segment the "grass", "house" and "sea" area of the image. The regions such as "grass" region, "house" region and "sea" region obtained from k-means algorithm for the "Satellite image 1" are shown in Fig. 7a–c respectively. Similarly, the regions such as "grass" region, "house" region and "sea region" obtained from k-means algorithm for the "Satellite image 2" are shown in Fig. 11a–c respectively. The SegNet is trained with the input images and with the ROI obtained from K-means algorithm and its label images.

Fig. 6 Input for "Satellite image 1"

(a) Grass region   (b) House region   (c) Sea region

Fig. 7 ROI obtained from k-means algorithm for "satellite image 1"

Fig. 8 Output obtained from SegNet when trained with "Satellite image 1" and its ROI obtained from k-means algorithm

Fig. 9 SegNet output when trained with "Satellite image 1" and its label image

Fig. 10 Input for "Satellite image 2"

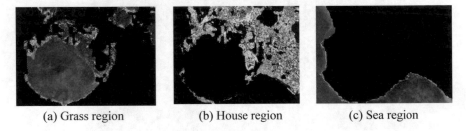

(a) Grass region   (b) House region   (c) Sea region

Fig. 11 ROI obtained from k-means algorithm for "satellite image 2"

In the SegNet, 4 encoder network and 4 decoder network is used. The convolution and deconvolution filter size is kept fixed to 3 × 3. The bias term in all convolutional layers are fixed to zero. A detailed description of the encoder–decoder layers is shown in Table 1. From the table the input to the Segnet is of size 350 × 467 × 3. The encoder1 performs convolution with 64 filter of size 3 × 3 with zero padding 1 and stride 1. The size of the output the convolution layer is 350 × 467 × 64. Then batch normalization is done on the convolution layer output [12]. After that element-wise ReLU operation, (max (0, x)) is performed. Following that, max-pooling with a non-overlapping window of size 2 × 2 and stride 2 is performed. The size of output of max-pooling is 175 × 233 × 64. Then in encoder2 convolution with 64 filter of size 3 × 3 with zero padding 1 and stride 1 is performed. The size of the output the convolution layer is 175 × 233 × 64. Then batch normalization is done. After that ReLU operation, (max (0, x)) is performed. Then max-pooling with a non-overlapping window of size 2 × 2 and stride 2 is performed. The size of output of max-pooling is 87 × 116 × 64. Then encoder3 performs convolution same filter size as in encoder 2 and encoder 3. The size of the output the convolution layer in encoder3 is 87 × 116 × 64. Then again batch normalization is performed. After that again element-wise ReLU operation performed. Then max-pooling is performed.

Table 1 Detailed description of encoder- decoder layers of the SegNet used for the present work

Encoder block			Decoder block		
Input		**Output size**	**Output**		**Output size**
Image input	350 × 467 × 3 images with 'zerocenter' normalization		Decoder 1 — Pixel classification layer	Cross-entropy loss with 'sea', 'house', and 1 other classes	
			Softmax	Softmax	350 × 467 × 64
Encoder 1 — Convolution	64 × 3 × 3 convolutions with padding 1 and stride 1	350 × 467 × 64	Convolution	64 × 3 × 3 convolutions with padding 1 and stride 1	350 × 467 × 64
Batch normalization	Batch normalization with 64 channels		Batch normalization	Batch normalization with 64 channels	
ReLU	ReLU		ReLU	ReLU	
Convolution	64 × 3 × 3 convolutions with padding 1 and stride 1	350 × 467 × 64	Convolution	64 × 3 × 3 convolutions with padding 1 and stride 1	350 × 467 × 64
Batch normalization	Batch normalization with 64 channels		Batch normalization	Batch normalization with 64 channels	
ReLU	ReLU		ReLU	ReLU	
Max pooling	2 × 2 max pooling with stride 2 and padding 0	175 × 233 × 64	Max unpooling	Max unpooling	350 × 467 × 64
Encoder 2 — Convolution	64 × 3 × 3 convolutions with padding 1 and stride 1	175 × 233 × 64	Decoder 2 — Convolution	64 × 3 × 3 convolutions with padding 1 and stride 1	175 × 233 × 64

(continued)

Table 1 (continued)

Encoder block			Output size	Decoder block			Output size
	Batch normalization	Batch normalization with 64 channels			Batch normalization	Batch normalization with 64 channels	
	ReLU	ReLU			ReLU	ReLU	
	Convolution	64 × 3 × 3 convolutions with padding 1 and stride 1	175 × 233 × 64		Convolution	64 × 3 × 3 convolutions with padding 1 and stride 1	175 × 233 × 64
	Batch normalization	Batch normalization with 64 channels			Batch normalization	Batch normalization with 64 channels	
	ReLU	ReLU			ReLU	ReLU	
	Max pooling	2 × 2 max pooling with stride 2 and padding 0	87 × 116 × 64		Max unpooling	Max unpooling	175 × 233 × 64
Encoder 3	Convolution	64 × 3 × 3 convolutions with padding 1 and stride 1	87 × 116 × 64	Decoder 3	Convolution	64 × 3 × 3 convolutions with padding 1 and stride 1	87 × 116 × 64
	Batch normalization	Batch normalization with 64 channels			Batch normalization	Batch normalization with 64 channels	
	ReLU	ReLU			ReLU	ReLU	
	Convolution	64 × 3 × 3 convolutions with padding 1 and stride 1	87 × 116 × 64		Convolution	64 × 3 × 3 convolutions with padding 1 and stride 1	87 × 116 × 64

(continued)

Table 1 (continued)

Encoder block			Output size	Decoder block			Output size
	Batch normalization	Batch normalization with 64 channels			Batch normalization	Batch normalization with 64 channels	
	ReLU	ReLU			ReLU	ReLU	
	Max pooling	2 × 2 max pooling with stride 2 and padding 0	43 × 58 × 64		Max unpooling	Max unpooling	87 × 116 × 64
Encoder 4	Convolution	64 × 3 × 3 convolutions with padding 1 and stride 1	43 × 58 × 64	Decoder 4	Convolution	64 × 3 × 3 convolutions with padding 1 and stride 1	43 × 58 × 64
	Batch normalization	Batch normalization with 64 channels			Batch normalization	Batch normalization with 64 channels	
	ReLU	ReLU			ReLU	ReLU	
	Convolution	64 × 3 × 3 convolutions with padding 1 and stride 1	43 × 58 × 64		Convolution	64 × 3 × 3 convolutions with padding 1 and stride 1	43 × 58 × 64
	Batch normalization	Batch normalization with 64 channels			Batch normalization	Batch normalization with 64 channels	
	ReLU	ReLU			ReLU	ReLU	
	Max pooling	2 × 2 max pooling with stride 2 and padding 0	21 × 29 × 64		Max unpooling	Max unpooling	43 × 58 × 64

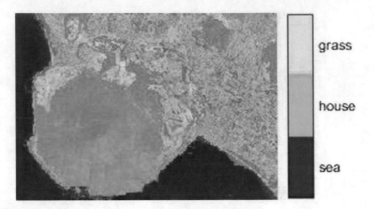

Fig. 12 Output obtained from SegNet when trained with "Satellite image 2" and its ROI obtained from k-means algorithm

Now, the size of output of max-pooling is $43 \times 58 \times 64$. Then in encoder4 the same operations are repeated. The size of the output in convolution layer is $43 \times 58 \times 64$ and maxpooling is $21 \times 29 \times 64$. The indices of max pooling are recorded in each encoder and used in the decoder network. The decoder network upsamples the input feature maps using the stored max-pooling indices which were recorded in each encoder network. From Table 1, the size of the input to the decoder4 is $21 \times 29 \times 64$. These feature maps are then upsampled and the size after unpooling is $43 \times 58 \times 64$. Then deconvolution is performed with 64 filter of size 3×3 with zero padding 1 and stride 1. The output size in this step is $43 \times 58 \times 64$. Then ReLU and batch normalisation is performed. Then Again in decoder3 the feature maps are upsampled and the size after unpooling is $87 \times 116 \times 64$. Then deconvolution is performed with 64 filter of size 3×3 with zero padding 1 and stride 1. The output size in this step is $87 \times 116 \times 64$. Then ReLU and batch normalisation is performed. Then in decoder2 again upsampling, deconvolution, batch normalization and ReLU are applied. The size after upsampling and deconvolution is $175 \times 233 \times 64$. Then in decoder1, after upsampling and deconvolution the size of output is $350 \times 467 \times 64$. Finally, the output of the final decoder is fed to a soft-max classifier which can be trained. This soft-max classifies each pixel independently. The predicted segmentation of the softmax is the class with maximum probability at each pixel. The weights of the Convolution layers in the encoder and decoder sub-networks are initialized using the 'MSRA' weight initialization method. The output of the proposed method is shown in Figs. 8 and 12. The method is also compared with the results of SegNet when trained with the "Satellite image 1" and "Satellite image 2" and its labeled images (Figs. 9 and 13). The accuracy of the method is also compared by increasing the depth of encoder to 6 and 8.

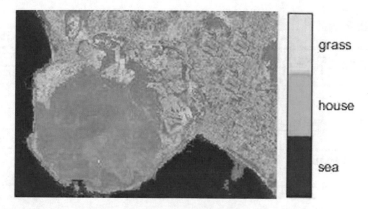

Fig. 13 SegNet output when trained with "Satellite image 2" and its label image

The experimental results are evaluate with the following metrics

a. Accuracy:

It indicates the percentage of correctly identified pixels for each class. For each class, the ratio of correctly classified pixels to the total number of pixels in that class is called the accuracy. Therefore,

$$Accuracy = TP/(TP + FN) \qquad (10)$$

Here, TP and FN are number of true positive and number of false negative respectively [33].

b. Intersection over union (IoU):

For each class, the ratio of correctly classified pixels to the total number of ground truth and predicted pixels in that class is called the IoU [33]. Therefore,

$$IoU = TP/(TP + FP + FN) \qquad (11)$$

where, TP is the number of True positive and FN is the number of false negative and FP is false positive.

c. Weighted IoU:

The average IoU of each class, weighted by the number of pixels in that class is called the weighted IoU [33].

Table 2 shows the experimental results obtained for "Satellite image 1". It was seen that the accuracy is 0.94356 for the proposed method and that for the SegNet is 0.93265. Again the mean IoU is 0.80774 for the proposed method and that for the SegNet is 0.80653. Finally the weighted IoU is 0.88297 for the proposed method and that for the SegNet is 0.86843.

Table 2 Experimental results for "satellite image 1"

Images	Accuracy	Mean IoU	Weighted IoU
Proposed method (Segnet + KMCA)	0.94356	0.80774	0.88297
SegNet	0.93265	0.80653	0.86843

Table 3 Experimental results for "satellite image 2"

Method	Accuracy	Mean IoU	Weighted IoU
Proposed method (Segnet + KMCA)	0.92635	0.80646	0.87864
SegNet	0.92454	0.80321	0.86432

Table 3 shows the experimental results obtained for "Satellite image 2". It was seen that the accuracy is 0.92635 for the proposed method and that for the SegNet is 0.92454. Again the mean IoU is 0.80646 for the proposed method and that for the SegNet is 0.80321. Finally the weighted IoU is 0.87864 for the proposed method and that for the SegNet is 0.86432. Thus for both the images the proposed method gives better result.

Tables 4 and 5 shows the how accuracy of the method changes with the increase in encoder depth. It was seen that when depth of encoder increases accuracy is also increases. In Figs. 14 and 15 graph between encoder depth versus accuracy is also plotted for both the images.

Figure 16 shows the normalized confusion matrix for the proposed method when trained with "satellite image 1". It is seen that 97.75% of the sea region, 98.61% of the house region and 87.25% of the grass region is predicted correctly. Thus the sea and the house region predicted properly than the grass region. In Fig. 17 the normalized confusion matrix for "satellite image 1" when trained with SegNet is shown. It is seen that 96.65% of the sea region, 99.61% of the house region and

Table 4 Experimental results when depth of encoder is increased for satellite image 1

Method	Encoder depth	Accuracy
The proposed method	4	0.94356
	6	0.95479
	8	0.95937

Table 5 Experimental results when depth of encoder is increased for satellite image 2

Method	Encoder depth	Accuracy
The proposed method	4	0.92635
	6	0.92889
	8	0.93437

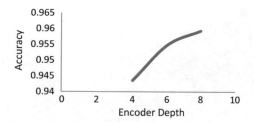

Fig. 14 Encoder depth versus accuracy plot for "satellite image 1"

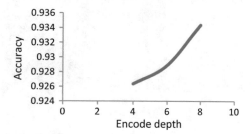

Fig. 15 Encoder depth versus accuracy plot for "satellite image 2"

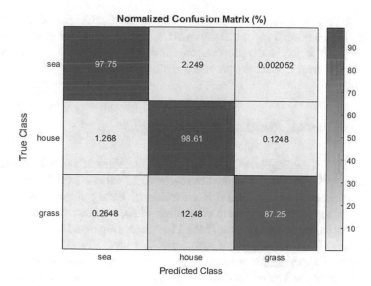

Fig. 16 Confusion matrix for SegNet when trained with "Satellite image 1" and its ROI obtained from k-means algorithm

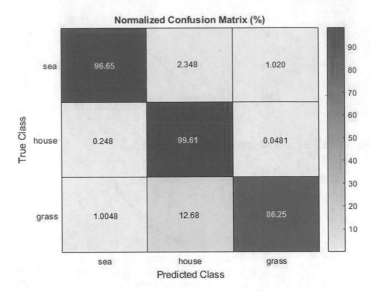

Fig. 17 Confusion matrix for SegNet when trained with "Satellite image 1" and its label image

86.25% of the grass region is predicted correctly by the SegNet method. Thus when compared the two normalized confusion matrix, it is seen that the prediction result for sea and grass region in the proposed method is better than the SegNet method. The training progress plot for the proposed method is shown in Fig. 18.

4.2 Experimental Results for the Classifier

With the output of the SegNet as input and the pixel values of various ROI as target a neuro-computing structure is trained to segment the various ROI. Here the neuro-computing structure is a MLP. The MLP is trained with BP algorithm. At first different training functions are used to train the classifier. Then, between the classifier output and target image, mean square error (MSE) had been calculated for all the training function and for "Sea" the regions of both "satellite image 1" and "satellite image 2" the images. The experimental results are shown in Tables 6 and 8. Among all the training functions the Levenberg-Marquardt (LM) back-propagation shows better MSE. The network is then trained with different number of hidden layers i.e. single, double and triple. The numbers of hidden neurons in hidden layers are altered and LM back-propagation training function is used to train the classifier. After that, for the images, MSE between the classifier output and target image is calculated. In Tables 7 and 9 the results are shown. In Figs. 19 and 20, the outputs of the classifier are shown.

At first classifier is trained using training functions such as LM Back-propagation, Resilient back-propagation, Conjugate Gradient back-propagation with polak-ribiere

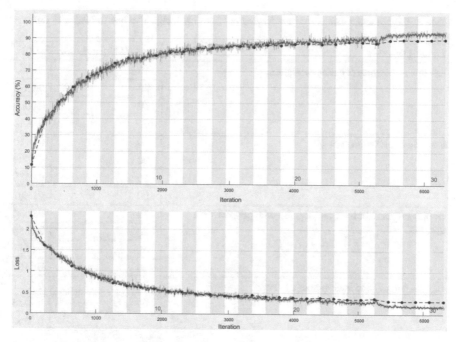

Fig. 18 Training progress plot with 6150 epoch and maximum iteration 6150

updates, Scaled conjugate Gradient, Bayesian Regulation back-propagation training functions. The hidden layer neuron numbers are taken as 10, 20, 60, 100. The results obtained are shown in Tables 4 and 6. It was seen that in terms of MSE LM Back-propagation algorithm shows better output. MSE is around 0.0165 when hidden neuron number 10 for "Satellite image 1". Again MSE is 0.0048 when hidden neuron was 100 units for "Satellite image 1". Similarly for "Satellite image 2" MSE is around 0.0559 when hidden neuron number 10. Again it is 0.0362 when hidden neuron was 100 units. The MSE value calculated for the other training functions, is much higher than LM Back-propagation. It was also seen that the values of MSE decreases when hidden neuron numbers increases,

Then in Tables 5 and 7 the results obtained when the classifier is trained with increased number of hidden layer i.e. double and triple are shown. Here LM Back-propagation is used as training algorithm. From the tables, the triple hidden layer shows better MSE values than double hidden layer and single hidden layer. For "Satellite image 1" the MSE for double hidden layer was around 0.0037 and MSE for triple hidden layer was calculated as around 0.0027. Similarly, the MSE for double hidden layer was around 0.0310 and MSE for triple layer was calculated as around 0.0273 for "Satellite image 2".

Table 6 Experimental results for "sea" region of "Satellite image 1" when trained with different training functions

Training function	No. of hidden neuron	MSE
Levenberg-Marquardt back-propagation	10	0.0165
	20	0.0118
	60	0.0059
	100	0.0048
Scaled conjugate gradient	10	0.0416
	20	0.0304
	60	0.0326
	100	0.0267
Conjugate gradient back-propagation with polak-ribiere updates	10	0.0634
	20	0.0538
	60	0.0574
	100	0.0452
Bayesian regulation back-propagation	10	0.0521
	20	0.0456
	60	0.0362
	100	0.0312
Resilient back-propagation	10	0.0624
	20	0.0563
	60	0.0547
	100	0.0512

Table 7 Experimental results for "sea" region of "satellite image 1" when trained with single, double and triple hidden layer

Classifier with different size of hidden layer	No. of hidden neuron	MSE
Classifier with single hidden layer	10	0.0165
	20	0.0118
	60	0.0059
	100	0.0048
Classifier with double hidden layer	[20 10]	0.0039
	[20 50]	0.0037
Classifier with triple hidden layer	[20 20 10]	0.0027
	[50 20 10]	0.0026

Table 8 Experimental results for "sea" region of "Satellite image 2" when trained with different training functions

Training function	No. of hidden neuron	MSE
Levenberg-Marquardt back-propagation	10	0.0559
	20	0.0441
	60	0.0437
	100	0.0362
Scaled conjugate gradient	10	0.0649
	20	0.0539
	50	0.0436
	100	0.0421
Conjugate gradient back-propagation with polak-ribiere updates	10	0.0613
	20	0.0605
	60	0.0593
	100	0.0582
Bayesian regulation back-propagation	10	0.0595
	20	0.0578
	60	0.0566
	100	0.0534
Resilient back-propagation	10	0.645
	60	0.0636
	50	0.0596
	100	0.0579

Table 9 Experimental results for "sea" region of "Satellite image 2" when trained with single, double and triple hidden layer

Classifier with different size of hidden layer	No. of hidden neuron	MSE
Classifier with single hidden layer	10	0.0559
	20	0.0441
	60	0.0437
	100	0.0362
Classifier with double hidden layer	[20 10]	0.0334
	[20 50]	0.0310
Classifier with triple hidden	[20 20 10]	0.0283
	[50 20 10]	0.0273

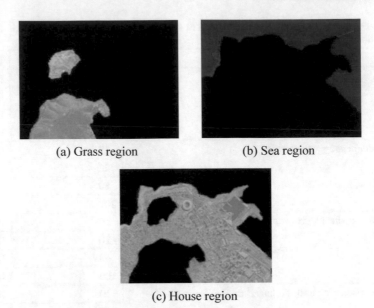

(a) Grass region   (b) Sea region

(c) House region

Fig. 19 Classifier output for "Satellite image 1"

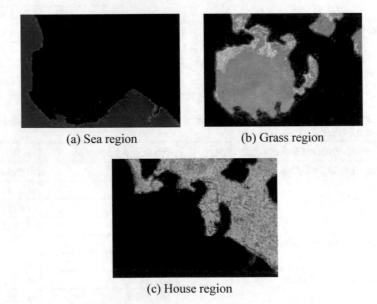

(a) Sea region   (b) Grass region

(c) House region

Fig. 20 Classifier output for "Satellite image 2"

5 Conclusion

In the chapter, some of the recent methods developed using deep learning based approaches for semantic segmentation are discussed. It includes the details of adopted deep learning topologies, experiments performed, results and related discussion. The works includes a specially configured and trained CNN topology which is found to be suitable for satellite image segmentation. Finally a method for semantic segmentation based on deep learning, in satellite images is also proposed. The proposed method comprises of the training and formation of a SegNet where the input images are satellite images. In the work, the target to the network was taken from the output of KMCA with their label of the required ROI. Then, with the output of the SegNet as input and the pixel values of various ROI as target a neuro-computing structure is trained to segment the various ROI. Here a MLP is used as the classifier which is trained with BP Algorithm. The approach does not use any pre-processing techniques such as region growing, region split-merge etc. to configure and train the SegNet, which is a deep convolutional network comprises Encoder-Decoder structure and the MLP. The method tested with different satellite images. The experimental results for both SegNet deep neural network and the classifier are shown. From the experimental results it was seen that the method is appropriate for real world and reliable.

References

1. Blaschke T (2010) Object based image analysis for remote sensing. ISPRS J 65:2–16
2. Salma EF, Mohammed EH, Mohamed R, Mohamed M (2016) A hybrid feature extraction for satellite image segmentation using statistical global and local feature. In: El Oualkadi A, Choubani F, El Moussati A (eds) Proceedings of the mediterranean conference on information & communication technologies 2015. Lecture notes in electrical engineering, vol 380. Springer, Cham
3. Sevak JS, Kapadia AD, Chavda JB, Shah A, Rahevar M (2017) Survey on semantic image segmentation techniques. In: International conference on intelligent sustainable systems (ICISS), Palladam, pp 306–313
4. Shotton J, Johnson M, Cipolla R (2008) Semantic Texton forests for image categorization and segmentation. In: Proceedings of IEEE conference on computer vision and pattern recognition, pp 1–6, Anchorage, AK, USA
5. Lin X, Wang X, Cui W (2014) An automatic image segmentation algorithm based on spiking neural network model. In: Proceedings of international conference on intelligent computing, pp 248–258. Springer, Taiyuan
6. Badrinarayanan V, Kendall A, Cipolla R (2017) SegNet: a deep convolutional encoder-decoder architecture for image segmentation. IEEE Trans Pattern Anal Mach Intell 39:2481–2495
7. Shotton J, Fitzgibbon A, Cook M, Sharp T, Finocchio M, Moore R, Kipman A, Blake A (2016) Real-time human pose recognition in parts from single depth images. In: CPVR, pp 1–8, Colorado Springs, CO, USA
8. Krizhevsky Krizhevsky A, Sutskever I, Hinton G (2012) Imagenet classification with deep convolutional neural networks. Adv Neural Inf Process Syst
9. Chen X, Xiang S, Liu C, Pan C (2013) Vehicle detection in satellite images by parallel deep convolutional neural networks. In: 2nd IAPR Asian conference on pattern recognition (ACPR), pp 181–185

10. Zhu H, Qi J (2011) Using genetic neural networks in image segmentation researching. In: Proceedings of IEEE international conference on multimedia technology (ICMT), pp 1–6, Hangzhou, China
11. Uijlings JRR, Sande VD, Gevers T, Smeulders AW (2013) Selective search for object recognition. Int J Comput Vision 104(2):154–171
12. Jiang Q, Cao L, Cheng M, Wang C, Li J (2015) Deep neural networks-based vehicle detection in satellite images. In: International symposium on bioelectronics and bioinformatics (ISBB), Beijing, pp 184–187
13. Long J, Shelhamer E, Darrell T (2015) Fully convolutional networks for semantic segmentation. IEEE conference on computer vision and pattern recognition (CVPR), Boston, MA, pp 3431–3440
14. Szegedy C, Liu W, Jia Y, Sermanet P, Reed S, Anguelov D, Erhan D, Vanhoucke V, Rabinovich A (2014) Going deeper with convolutions. CoRR, abs/1409.4842
15. Simonyan K, Zisserman A (2014) Very deep convolutional networks for large-scale image recognition. CoRR, abs/1409.1556
16. Krizhevsky A, Sutskever I, Hinton GE (2012) Imagenet: classification with deep convolutional neural networks. In: NIPS
17. Donahue J, Jia Y, Vinyals O, Hoffman J, Zhang N, Tzeng E, Darrell T (2014) DeCAF: a deep convolutional activation feature for generic visual recognition. In: ICML
18. Ronneberger O, Fischer P, Brox T (2015) U-Net: convolutional networks for biomedical image segmentation, computer science department and BIOSS centre for biological signalling studies, University of Freiburg, Germany
19. Yoshihara A, Takiguchi T, Ariki Y (2017) Feature extraction and classification of multispectral imagery by using convolutional neural network. In: International workshop on frontiers of computer vision
20. Jégou S, Drozdzal M, Vazquez D, Romero A, Bengio Y (2017) The one hundred layers tiramisu: fully convolutional DenceNets for semantic segmentation
21. Patravali J, Jain S, Chilamkurthy S (2018) 2D-3D fully convolutional neural networks for cardiac MR segmentation. In: Pop M et al (eds) Statistical atlases and computational models of the heart. ACDC and MMWHS challenges. STACOM 2017. Lecture notes in computer science, vol 10663. Springer, Cham
22. Langkvist M, Kiselev A, Alirezaie M, Loutfi A (2016) Classification and segmentation of satellite orthoimagery using convolutional neural networks. Remote Sens 8(329)
23. Huang G, Liu Z, Weinberger KQ, Maaten LV (2016) Densely connected convolutional networks. CoRR, abs/1608.06993
24. Chaurasia A, Culurciello E (2017) LinkNet: exploiting encoder representations for efficient semantic segmentation
25. He K, Gkioxari G, Dollar P, Girshick R (2017) Mask R-CNN
26. United States Geological Survey. https://earthexplorer.usgs.gov/
27. Maggiori E, Tarabalka Y, Charpiat G, Alliez P (2017) Can semantic labeling methods generalize to any city? The inria aerial image labeling benchmark. In: IEEE international geoscience and remote sensing symposium (IGARSS), Fort Worth, TX, pp 3226–3229
28. DeepGlobe CVPR 2018—Satellite Challenge. http://deepglobe.org/
29. Barthakur M, Sarma KK (2018) Complex image segmentation using K-means clustering aided neuro-computing. In: 5th international conference on signal processing and integrated networks (SPIN), Noida, Delhi
30. Liu Y, Minh Nguyen D, Deligiannis N, Ding W, Munteanu A (2017) Hourglass-shape network based semantic segmentation for high resolution aerial imagery. Remote Sens 9:522
31. Yasrab R, Gu N, Zhang X (2017) An encoder-decoder based convolution neural network (CNN) for future advanced driver assistance system (ADAS). Appl Sci 7:312

32. Haykin S (2003) Neural networks a comprehensive foundation, 2nd edn. Pearson Education, New Delhi
33. Barthakur M, Sarma KK, Mastorakis N (2017) Learning aided structures for image segmentation in complex background. In: 2017 European conference on electrical engineering and computer science (EECS)

A Deep Dive into Supervised Extractive and Abstractive Summarization from Text

Monalisa Dey and Dipankar Das

Abstract Since the advent of World Wide Web, the world has seen an exponential growth in the information on the internet. In the web, today, a lot of digital articles are available for users to read. However, these documents lack a concise summary. Automatic systems that summarize such information are very rare. Moreover, these systems also lack the ability of getting information from multiple places and presenting the user with a coherent, informative summary. Summarization are of two types, viz. Extractive and Abstractive. While a lot of work has been done on extractive summary, abstractive summary generation is still an unexplored area. In the current work, the focus is essentially on producing abstractive summary of a document by first getting relevant and important sentences from the corpus and feeding these sentences as input in a trained deep neural network. For extracting salient sentences two models has been proposed, both these models not only consider syntactic similarity but also semantic similarity between sentences. To capture semantic similarity, embedding models like word2vec or paragraph vectors has been used. Next, to generate abstract version of these sentences a RNN autoencoder is used, which is responsible for the representation of longer sentences into shorter sentences without loosing semantic information. ROUGE evaluation technique has been used for evaluating the generated summary quality.

Keywords Extractive · Abstractive · Semantic similarity · Deep neural network

M. Dey (✉) · D. Das
Department of Computer Science and Engineering, Jadavpur University,
Kolkata, India
e-mail: monalisa.dey.21@gmail.com

D. Das
e-mail: dipankar.dipnil2005@gmail.com

© Springer Nature Switzerland AG 2020
J. Hemanth et al. (eds.), *Data Visualization and Knowledge Engineering*,
Lecture Notes on Data Engineering and Communications Technologies 32,
https://doi.org/10.1007/978-3-030-25797-2_5

1 Introduction

Natural language processing, a field under the umbrella of Artificial Intelligence and Computer Linguistics attempts for reducing man machine gap. In this area, researchers are trying to build the systems that will have the capability to process and understand huge amount of data in natural languages. The main output of a summarization system is to generate shorter summarized texts for normal texts. A high quality summary should contain all the informative parts of the source document(s)) besides being coherent, concise and correct grammatically.

One of the basic distinguishing parameter among different types of summarization is based upon the input texts. A summary can be formed from a single source or from more than one source as input. Summary from a single document is mainly used for generating simple headline outputs, outlines of a text or snippet generation. The objective for summarizing more than one document, is to condense all the text available in the documents and to generate a summary that encompasses the relevant information present in all the documents. One of the most widely used distinguishing feature is, the manner in which the summarization process is implemented. An extractive summarizer works by selecting exact sentences from the source document(s) and including the most important ones in the summary; the material usually being sentences, paragraphs or even phrases. On the other hand, abstraction based summarization does not rely on simple extraction to generate summary; instead it works by creating new sentences intelligently from the given document(s). Summarization systems can work on text or multimedia. In this work the main focus is on text summarization. One significant challenge in developing a summarization system is differentiating the most relevant and important parts of a document from the unimportant ones.

Although there are a few works on generating abstractive versions of a summary, most of the work done is in extractive summarization on single source texts. One other problem in summarization is that if two different human beings are given the same source text, they will form two completely different summaries based on their intelligence and knowledge. One of the most important and challenging problem in case of abstractive summarization is natural language generation. In single document summarization, one of the earlier implementation deals with compression and fusion techniques to generate new sentence, both these techniques are less likely to produce grammatically correct and readable sentences. Summarization also possesses challenge in the evaluation of summary output generated by the system. One good way to do it is by asking humans to evaluate the summary. Another way might be to ask human beings to write their own versions of summary and then compare them to the system's output. In order to automate this, a set of metrics known as Rouge has been designed. However, the process of evaluation has still a long way to go.

This work aims to propose modules to form abstractive summary of a given source corpus. The primary work is divided into two parts mainly, first generate the extractive summary of the document and try to convert this extractive summary into abstractive summary using deep neural network. In the very first part there are some novel techniques of content selection, which will not only consider the lexical structure of words in the text but also takes into account its semantic structure and syntactic structure by sing word embedding or sentence embedding and pos tagging. The objective of second part is mostly to generate new and shorter sentences to a given sentence, which is required to generate complete abstractive summary. While generating new sentence for a given sentence it removes the clutters from it and gives attention to important words in the input.

2 Literature Survey

Since the initial study by Luhn in 1958, a lot of researchers have studied this area during the 1990s, influenced by conferences like TREC and DUC. Summarization methods are of two different types: Structured and semantic approaches.

3 Structure Based Methods

3.1 Tree Structure Based Approach

This is dependency tree based approach [3]. Initially, sentences which are alike are preprocessed and mapped to an argument type structure. Next, theme intersection algorithm is used to find similar phrases by comparing these structures. After that, those phrases which kind of conveys same meaning are selected and some extra information like temporal reference, newswire source references and entity description are augmented. At the end, these selected phrases are arranged as the new summary using algorithms like SURGE/FUR. One disadvantage of this approach is the generated summary does not take into account the contextual meaning of the sentences.

3.2 Rule Based Method

The rule based extraction module used by this method [12], the content selection analytics and sentence creation patterns are all hand written. Text generation is implemented using the SimpleNLG realizer, whose input is a sentence with each words root from and output is a sentence with markers.

One drawback of using this method is the effort taken to manually write all the rules.

4 Semantic Approaches

4.1 Multimodal Semantic Model

In this work, [13], the authors created a semantic model, capturing concepts and their relationships, which are then used to build to form the content of the document. The most significant concepts are identified and then used to create the summary of the text.

4.2 Information Item Based Method

In this, the authors [11] have created the summary from conceptual form of the source texts and not directly from the source. This conceptual and abstract form is known as an information item which represents the smallest unit of the summary. Their model consists of the following parts: formation of information items, generating sentences and forming the final summary. On major disadvantage is the quality of the generated output is very low.

4.3 Semantic Graph

The aim of this method [24] is to create a graph from the given source text known as the semantic graph by reducing the original semantic graph of the source corpus. The final output is then generated from this rich graph. In a related work, [21] authors have used a model which creates AMR graphs. These are then converted into a final summary graph. A drawback of this approach is that it is a generalized approach and not specific to a domain.

5 Other Methodologies

5.1 Content Based Models

Some of these approaches are explained below:

5.1.1 Term Models

This type of content frameworks, give scores to sentences based on their importance by following some statistic model which work on distribution of words. The initial models, focused on finding important words from the document depending on the number of times these words appear in the document. Later however, Edmundson [7] used a linear combination of numerous scores based on whether these words are cue words, title or head words. He also used the location of the sentence in the document as an important parameter while ranking sentences. Kupiec [16] treated this as a classification problem and used the most legendary Naive Bayes classifier by training it with features like frequency of words, position of a sentence, length and also presence of capital phrases. The data that was used for training, came from scientific articles and also abstracts that were manually written. The sentences in these selected abstracts were aligned with the sentences in the source document for creating data for training extractive models. A very well known model known as SUMMARIST [20] was created by Lin and Hovy which used topic signature for summarization. Generally, a family of related terms, which are also significantly pointing towards a single concept are known as topic signatures. Topic signatures are collected from a pre-classified document and are then ranked using a famous test known as the likelihood ratio test. In this, the score assigned to each of the sentences are nothing but the summation of all the content words in that sentence. Another similar structure [28] which has been studied for modelling important words is known as a centroid. It is defined as a group of words which bear statistical importance to a set of text documents. The model then gives a centroid based score to the sentences which is also based on the position and overlap of that sentence with the first sentence.

5.1.2 Network Models

The model described above, does not take into account the similarity index between sentences of the same document. Networks might be a better and improved way to model such relationships. These approaches [8] work by creating a network out of the sentences in the source document. Thus, sentences become nodes, and edges between them are weighed depending on how much they are similar to each other. An algorithm known as PageRank is then used to identify relevant sentences. Another such method known as C-Lexrank [27] does a few changes in the original LexRank by initially breaking up the networks into sub-parts based on its communities and then identify relevant sentences from each of these sub networks.

5.1.3 Probabilistic Models

These models are used mainly for summarizing multiple documents and mainly follow Bayesian word distribution technique. An early model is BayeSum [6] which mainly focuses on query dependent summarization. Another such model is TopicSum [15] where the query based model is substituted by a language model which is specific to a collection of documents. In this approach, the importance of a sentence depends on the fact that how likely is it that it will contain information regarding the entire collection of document rather than a single document. The more likely it is, the greater the score the sentence will get. Authors here have also introduced a more sophisticated version known as HierSum, which divides this collection of documents into multiple topic and contents which represent the variety of content the documents might be about. In a similar kind of work, Fung and Ngai in 2006, presented an HMM model. However, this time it was for summarizing multiple documents [9].

5.2   Generation Model: NAMAS

This model [30] takes x as an input which is a sequence of M words $x_1, x_2, x_3, \ldots, x_M$. The output is a more reduced version of the sentence y which is of length $N < M$ where y is itself sequence of words $y_1, y_2, y_3, \ldots, y_N$. Every word, in the source input and the output summary belongs to a fixed vocabulary $v = |V|$. It assumes that the length of the output produced is not variable. The system also knows the summary length prior to the generation. The system tries to find the optimal sequence from all the possible sequences of length N. This model uses the attention based encoder, that is discussed in detail.

This model focuses on scoring function s which considers a window of fixed length previous words.

$$s = \sum g(y_{i+1}, x, y_c)$$

where $y_c = y_{i-c+1,\ldots,i}$.

The main aim of this approach is to calculate $y^* = argmax_{y \in Y} \sum g(y_{i+1}, x, y_c)$, To decide on the most optimal value of y^*, a decoder known as beam search is used which maintains a an entire vocabulary while only using K hypothesis at each summary position.

Recent follow-up [25] of this model addresses some of the limitations of NAMAS.

1. This new model [25] uses the large vocabulary trick [23]. For each mini batch, its decoder vocabulary list only contains the words of that particular batch. The words which are more frequently used are kept on adding until it reaches that target size. The computation of the softmax layer in the decoder is reduced to get rid of the bottleneck of large sized vocabulary.
2. But then it loses the capability to do "abstractive" summary—which means introducing words which were not in the source—so they do "vocabulary expansion". This is done by augmenting a layer known as "word2vec nearest neighbours" along with words in the input.
3. A significant feature of this work is the usage of both switching generator and another pointer layer [33].
4. This model uses the concept of hierarchical Encoder with hierarchical attention. For the work, two RNN, both of which are bi-directional is used at source. First one is used for words and the second one for sentences.

6 Related Models

6.1 Extractive to Abstractive Using Sentence Compression and Fusion

This method was proposed by Fei Liu and Yang Liu, states that a good quality compression technique applied on extractive summary, document can improve the readability of the summary and make it more like an abstractive summary [22]. It uses different automatic compression algorithms. The compression guideline is described in Clarke and Lapata [5]. The employed external annotators were instructed to try and eliminate those words which do not change the semantic meaning of a sentence. They were also asked to try and preserve the grammatical structure and meaning.

6.2 Neural Networks and Third Party Features

In the year 2002, DUC arranged a task for summarization where the participants had to create a summary of 100 words for a news article. Some of the best systems submitted however, could not perform better than the baseline system. Svore [32] proposed a neural net based algorithm which used 3rd party data sets for tackling the issue of creating an extractive summary. RankNet [4] was used for ranking purposes which is a neural network algorithm which uses gradient descent for training the model. ROUGE-1 [19] was used for scoring the similarity between the output and the human written version. These scores were used as labels in contrast to other proposed approaches where sentences are hard labelled depending on they were selected or not.

7 Extractive Summarization

Extractive summarization deals with the selection of salient sentences from a document without changing the lexical structure of any sentence. Content selection is one of the important aspect of extractive summarization. One of main methods is text rank inspired from page rank [26], that can be used for ranking the sentences of a document according to its importance. For extracting the sentences, initially an entire sentence is ranked and for each of the sentences in the text, a vertex is added to the graph. We are trying to define a relationship which checks to see if there is any similarity between two sentences. Various features are used to get this similarity score like overlap of contents, over of their POS and the similarity score between the vector representation of the sentences.

In the current work, a normalization factor is also introduced in order to avoid inclusion of sentences which are beyond a certain length. This has been done to keep the similarity score between 0 and 1.

For finding similarity between sentences their vector representation has been considered. The methods used for sentence vector generation are simple weighted averaging of word vector obtained from google news word2vec module or glove pretrained vectors and then removing common component using PCA and doc2vec model trained from wikipedia corpus.

7.1 Word2vec and Weighted Average Word Vectors: Smoothing Invariance Factor

One of the most efficient models for understanding and learning the word embeddings from raw text. There are two types of this approach. First is the Continuous Bag of Words model and second is the skip gram model. When we take a look at the algorithms of these above mentioned types, they are similar except that the former predicts the target words from the context (e.g. 'table') from the source ('the dog is sitting on the'), whereas the later tries to predict the context from the target. Here a newer supervised approach will be discussed for sentence vector generation, the algorithm is pretty simple:

Algorithm: Sentence Embedding

Input: Word embeddings $v_w - w \in V$, a set of sentences S, parameter a and estimated probabilities $p(w) - w \in V$ of the words.
Output: Sentence embeddings $v_s - s \in S$

1. **for all** sentence s in S do
2. $\quad v_s \longleftarrow \frac{1}{s} \sum_{w \in s} \frac{a}{a + p(w)} v_w$
3. **end for**
4. Compute the first principal component u of $v_s - s \in S$
5. **for all** sentence s in S do
6. $\quad v_s \longleftarrow v_s - uu^T v_s$
7. **end for**

7.1.1 Paragraph Vectors

This section discusses about CBOW[1] and how it is trained and how paragraph vector is just an extension of it.

$$h = \frac{(w_1 + w_2 + \cdots + w_c)}{c} \tag{1}$$

---

[1]Continuous bag of words.

The update functions remain the same except that for the update of the input vectors, It is needed to apply the update to each word in the context C

$$W_{w_I}^{(t)} = W_{w_I}^{(t-1)} - \frac{1}{C} \cdot \nu \cdot EH \tag{2}$$

More formally as mentioned in "Distributed Representations of Sentences and Documents" [18], a sequence of training words are given, the aim of the model is maximizing avg. log probability.

$$\frac{1}{T} \sum_{t=k}^{T-k} log \ p(w_t \mid w_{t-k}, \ldots, w_{t+k}) \tag{3}$$

A multiclass classifier softmax does the prediction. There, we have:

$$p(w_t \mid w_{t-k}, \ldots, w_{t+k}) = \frac{\exp(yw_t)}{\sum_i \exp(y_i)} \tag{4}$$

y_i is the non-normalized log-probability for the output words i, which are computed as

$$y = b + Uh(w_{t-k}, \ldots, w_{t+k}; W) \tag{5}$$

where U, b are the softmax parameters. h is constructed by a concatenating or averaging word vectors which are extracted from W.

The approach used for learning paragraph vectors is inspired by the methods used for learning word vectors.

The Paragraph Vector model attempts to learn fixed-length continuous representations from variable-length pieces of text. These representations combine bag-of-words features with word semantics and can be used in all kinds of NLP applications.

Paragraph Vector is only a very small extension of the original model. Similar to the Word2Vec model, the Paragraph Vector model also predicts the next word that might come up in a sentence. The only real difference, as the paper and in the figure[2] itself is with the computation of h. Where in the original model h is based solely on W, in the new model another matrix called D is added, representing the vectors of paragraphs:

A paragraph is assumed to be another word. The difference is that, in this work, all the vector representation of the paragraphs are unique but the words share their vectors among different contexts. At each step we compute h by concatenating or averaging a paragraph vector d with a context of word vectors C:

$$h = \frac{(D_d + w_1 + w_2 + \cdots + w_c)}{C} \tag{6}$$

---

[2]https://nbviewer.jupyter.org/github/fbkarsdorp/doc2vec/blob/master/doc2vec.ipynb.

The weight update functions are the same as in Word2Vec except that the paragraph vectors are also updated. This first model is called the Distributed Memory Model of Paragraph Vectors. Le and Mikolov present another model called Distributed Bag of Words Model of Paragraph Vector. This model ignores the context C and attempts to predict a randomly sampled word from a randomly sampled context window.

Stochastic gradient descent is used for training both the word and paragraph vectors. The gradient is obtained using the process of back propagation. A fixed length context from any random paragraph is sampled during every step of the stochastic gradient descent. Next, the error is calculated and the parameters are updated accordingly. During predicting, an inference step is done for computing the vector of the new paragraph. This can also be obtained from gradient descent. The others parameters such as W (word token vectors), softmax weights are kept fixed. So, to sum up, there are two stages; (1) training to get the word token vectors W, the softmax weights U, b and finally D which are the paragraph vector representations (of the paragraphs which are already seen). (2) Paragraph vectors of the never before seen paragraphs after adding more columns in D and gradient descending on D while the values of W, U, b remain fixed. Logistic regression is used to predict about specific labels.

7.2 Page Rank

Google utilizes an algorithm known as PageRank [26] for ranking the web pages corresponding to a particular web query by a user. This algorithm is extremely efficient in analyzing directed graphs and not just web graphs. In the current work, this standard algorithm has been used for selection of content.

Node Scoring

1. The nodes of the graph, distributes the score amongst themselves. Suppose a node is pointing at 3 nodes (directed graph), then that node contributes one third of its score to each of the node it is pointing to.
2. To calculate the score of a single node, scores of all the nodes which points to it are summed up.

Let us see how easily this can be represented using linear algebraic terms. An adjacency matrix M is needed for representing the connections between nodes. A column vector 'r' will contain the scores of the nodes. We put a fraction: it points to, to represent the connections. The equation defining the relation between node scores is as given below:

$$\mathbf{r} = \mathbf{M} * \mathbf{r} \tag{7}$$

The form of the node score equation is similar to an equation of an eigen vector x:

$$A * x = \lambda * x \tag{8}$$

The score vector r can be interpreted as the eigenvector of the adjacency matrix M with an eigenvalue of $\lambda = 1$. Due to some properties of matrix M predefined by us, the first eigen vector of M ('r') will always hold the value 1 and will be the first eigen vector. A more simpler way of finding it is known as power iteration. In this, 'r' will be initialized with by 1/total no. of nodes. Then keep on calculating $r = M * r$ until it converges.

There is a distinct disadvantage to the PageRank algorithm that has been studied so far. There are two conditions that has to be satisfied for it to run flawlessly. First, M has to be stochastic and second, the graph must not contain any cycles. The example given in the paper may satisfy these but a graph for web pages will most definitely not.

In this work, to fix this problem the graph is tweaked a bit. A complete graph is created using all the web pages (there is a link between all the nodes), and the links are assigned very small weights. Algebraically the tweak can be represented as shown below:

$$A = \beta M + (1 - \beta)\frac{1}{n}e * e^{T} \tag{9}$$

M is the original matrix and A is the tweaked one.

7.3 Proposed Models for Extractive Summary Generation

7.3.1 Model 1: Utilizing Word2vec and Weighted Average Word Vectors

The very first model uses the inbuilt Word2vec model trained on Google news corpus or Glove[3] vectors trained on wikipedia2014 and Gigaword corpus, for sentence vector generation. These word vectors are weighted averaged using the estimated word frequency as discussed above, then sentence vectors are generated under smoothing invariance factor scheme in Sect. 7.1.

7.3.2 Model 2: Utilizing Paragraph Vectors

Both the Models 1 and 2 are different from each other as Model 2 does not consider weighted average of word vectors for sentence generation, rather it generates word vectors using paragraph vector. So in Model 2 similarity scores between sentences is calculated from the normalized values of correlation similarity between the sentence vectors obtained from paragraph vectors, word overlap and pos tags overlap.

---

[3]https://nlp.stanford.edu/projects/glove/.

8　Abstractive Summarization

8.1　Sentence Generation Using Deep Neural Network

Since for generating a semantically shorter sequence for a corresponding longer sequence, one need to understand the longer sequence first and then use this information every time while generating each word of the output sequence. A deep neural network possesses the ability of understanding and generating time dependent sequences. A deep neural network is typically designed as feedforward networks, but RNN and LSTM's are very frequently and successfully used for language modelling purposes.

For data which is sequential, RNN's are particularly helpful. This is so because, in this type of network, every neuron has its own state (memory). A recurrent neural network is specially useful with sequential data because each neuron or unit can use its internal memory to maintain information about the previous input.

Using RNN's can actually help in more accurate predictions of a word at the end of a sentence. But RNN actually suffers from the problem known as vanishing gradient, hence rather than using RNN LSTM is preferred because it has the ability to remember only a portion of previous context which further helps it in remembering longer sequences.

Now in next section we will see how abstractive summarization method is implemented using general sequence to sequence framework encoder- decoder RNN. We have drawn the inspiration of this model from encoder-decoder model proposed by Bahdanau [2]. This model will also use the same attention based encoder that learns a soft alignment over the input text to help inform the summary. But before going into some detailed overview of this model, we will come to know what sequence to sequence framework is and how it is used in conjugation with attention models. We will dig deeper further to see some concepts that will lead to the extension of the model.

8.2　Long Short Term Memory Network

RNN uses sequential information. RNN's are recurrent as they perform same function on every element of a sequence where the output id dependent on previous computations. RNN's also have memory for capturing information about what has been calculated so far. RNN is governed by the following formula:

- x_t is the input at time step t. For example, x_1 could be a one-hot vector corresponding to the second word of a sentence.
- s_t is the hidden state at time step t. Its the memory of the network. s_t is calculated based on the previous hidden state and the input at the current step: $s_t = f(Ux_t + ws_{t-1})$. The function f usually is a non linearity such as tanh or Relu. s_{-1} which is typically required to calculate the first hidden state is initialized to all zeros.

– o_t is the output at step t. For example, if we wanted to predict the next word in a sentence it would be a vector of probabilities across in the vocabulary $o_t = softmax(V s_t)$.

But a general vanilla RNN suffers from a problem which is known as vanishing gradient which prevents it from learning long term dependency. It will be discussed in the next section.

Gradient Problem Vanishing gradient is an issue observed during training a neural network with gradient based learning methods (for example, backpropagation). During updation of weights, the network receives updates by calculating the gradient of the error function with respect to the current weight. During the gradient computation, the gradients of activation functions like Sigmoid and Tanh squashes the values in a tiny restricted range $(0, 1)$ and $(-1, 1)$, respectively. In case of a recurrent neural network with n time steps, repeated multiplication of these n small gradients result in an infinitesimally small gradient i.e. the gradient "vanishes" which does not help the network learn anything.

Lstm's are used for combating vanishing gradients by the use of a gating mechanism. To understand what this means, the following section explains how a hidden state s_t is calculated by LSTM.

$$i = \sigma(x_t U^i + s_{t-1} W^i)$$
$$f = \sigma(x_t U^f + s_{t-1} W^f)$$
$$o = \sigma(x_t U^o + s_{t-1} W^o)$$
$$g = tanh(x_t U^g \mid s_{t-1} W^g)$$
$$c_t = c_{t-1} \circ f + g \circ i$$
$$s_t = \tanh(c_t) \circ o$$

A LSTM layer is another way of calculating a hidden state. Earlier, it was calculated as $s_t = \tanh(U x_t + W s_{t-1})$. In this the inputs were, x_t, which were the current input at step t, and s_{t-1}, which was the previous hidden state. s_t was the output hidden state. Let us see how LSTM does it.

– i, f, o are the input, forget and output gates, respectively. These have same equations defining them except for the parameter matrices. The amount of newly calculated state, amount of previous state and the amount of internal state that you want to let through is defined by the input gate, forget gate and output gate respectively. The dimension of all the gates are same and implied by d_s.
– g and c_t denotes the candidate hidden state and internal memory respectively. The candidate hidden state is calculated using the current input and previous hidden state. The previous memory, c_{t-1} is multiplied by the forget gate. The newly calculated hidden state g is multiplied with the input gate. And both of these together make up the internal memory of the unit c_t.
– The output hidden state s_t is at last computed by multiplying c_t with output gate.

RNNs are basically a special case of LSTMs. That means, that if the input gate value is fixed as 1's and the forget gate values are fixed as 0's, the previous memory is always forgotten and the output gate are all 1's thereby making it a standard RNN. LSTM's are however used mainly as they are capable of modeling long-term dependencies.

Deep neural nets are powerful models that excelled in much difficult learning tasks. Although it works well when a lot of labelled datasets are available. However, they cannot be used for mapping sequence to sequence. This issue can be very well handled by LSTM. The LSTM will read the input sequence at each timestamp and then a vector representation can be obtained which is large and fixed. Another LSTM then can be used for extracting the output sequence form that fixed input vector. Thus, this gives rise to an encoder-decoder model.

8.3   RNN Encoder Decoder

As already mentioned in Sect. 8.2 that a recurrent neural network consist of a hidden state s_t, optional output o operating on a variable length sequence $x = (x_1, x_2, x_3, \ldots, x_T)$. The hidden[4] state s_t is updated by

$$s_t = f(s_{t-1}, x_t) \tag{10}$$

where f is a non-linear activation function. This f can be a logistic sigmoid function calculated element wise, or a complex one like Sect. 8.2.

This architecture [31] learns to encode a variable-length sequence into a fixed-length vector representation. It is also capable of decoding a fixed-length vector representation into a variable-length sequence. The encoder is an RNN as described in the previous section. It reads the input symbol x sequentially and changes the hidden state as mentioned in (10). The decoder is again an RNN, which will generate the output sequence by predicting the next symbol o_t given the hidden state s_t or (h_t'). The hidden state of the decoder at time t is computed by,

$$s_t = f(s_{t-1}, y_{t-1}, h) \tag{11}$$

8.4   Neural Attention Model

In the encoder-decoder architecture described above, the vector is supposed to encode all the information of the source which is very unreasonable. In the attention mechanism, instead of encoding the entire sentence into a single vector, the decoder is

---

[4]In some context s_t and o is also denoted by h_t and Y respectively.

allowed to attend different parts of a source sentence at each step of the output generation.

The attention model addresses the limitation as described:

1. How to inform the decoder know of the part of encoding which is relevant at each step of the generation process.
2. Encoder has limited memory so there has to be a way to overcome this.

Thus, the attention mechanism, helps the decoder to pick only those encoded inputs that are important for the decoding process.

8.5 Sampled Softmax and Output Projection

Let us say we have a single-label problem. (x_i, t_i) is a training sample consisting of a context and a target class. $P(y \mid x)$ is the probability that target class is y given the context is x. We are aiming to train F(x, y) to produce softmax logits, which is relative log probabilities of the class given the context.

$F(x, y) \leftarrow log(P(y \mid x)) + K(x)$ K(x) is an arbitrary function that does not depend on y. In full softmax training, for every training example (x_i, y_i), we would need to compute logits $F(x_i, y)$ for all classes in $y \in L$. This can get expensive if the Universe of classes L is very large. In "Sampled Softmax",[5] for each training example (x_i, t_i), A small set $S\_i \subset L$ of "sampled" classes according to a chosen sampling function $Q(yx)$.

8.6 Handling Rare Words

The above-desired model has a number of advantages but its limitation is in vocabularies which have a very large size. This not only increases the training complexity but also the decoding complexity. Generally, the training corpus is partitioned and a subset of the target vocabulary is defined for each of these partitions. Prior to training, each target sentence in the source text is analyzed and unique words are collected upto a threshold. The collected unique word vocabulary will be used for this partition of the corpus during training. This is repeated until the end of the set.

One other pretty simple approach to address the rare word problem [23]. According to this model, if the source word responsible to each unknown word in the target sentence is known, we can introduce a post-processing step that could replace each unknown token from its corresponding word in the dictionary.

---

[5]https://www.tensorflow.org/extras/candidate_sampling.pdf.

8.7   Beam Search Decoder

Let us consider a basic seq2seq[6] with just one hidden layer and vanilla RNN for both encoder and decoder. Consider what happens when the encoding is done. The final hidden state is obtained from the encoder end RNN. Now the decoding process starts. The obtained final hidden state of the encoder is fed as the initial state of the decoding end RNN.

The end token of the input sequence is fed to the decoder input and output di is obtained which becomes the first word in the output sequence. And if this first word is given to the decoder input in the 2nd time step and you get the 2nd word in the output sequence. The process continues until you get an end token for the output sequence.

A particular sentence is obtained as the output of the decoder (which is the best possible output for the corresponding input).

Why the approach does not work: In the above approach, the assumption is that the best output sentence always start with di which is not a valid assumption. It might happen, a better sentence would have been the one that starts with the word that had the 2nd highest probability when the output is obtained in the first time step of the decoder above (jurisdiction needed).

Beam search decoder [34]: Rather than just considering the highest probable word as the input of the output sequence, the top b words is taken as the input of our decoder sentence. Here b is also called the beam size which is itself a parameter. Now to compute the second word in the decoder each of these b first words is fed and all the prefixes of length 2 are obtained (We get b*N sequences if N is the vocab size). Now for each first word out of the top b words, the probabilities of all the N words are obtained in the for the 2nd time step. So, out of the bN prefixes of length 2, chose the top b ones. Then predict the 3rd word conditioned on these b prefixes (of length 2) and again bN prefixes of length 3 are obtained and the top b ones are chosen. The process continues till the end. In the end we have top b decoded sequences as our output. This kind of looks like a greedy decoding process.

9   Experiment and Results

As I already mentioned all the underlying concepts of our experiment, in this section it can be seen how all the above concepts has been used in the experiments.

---

[6]Sequence to sequence.

10 Extractive Summary Generation

10.1 Dataset Used

For the current work, the Opinosis Dataset—Topic related review sentences' [10] has been used. A brief description of the dataset is given in Table 1. This dataset contained topics and their summaries. There were 51 topics having approx 108 sentences. The gold standard summaries have a set of 5 summaries generated by humans for each source text.

10.1.1 Preprocessing

Every sentence in each document is tokenized, stemmed using porter stemmer of nltk, later they are lemmatized. The non ascii character and string punctuations are removed from every sentences.

Both for noun and verb forms lemmatization has been done. Actually nltk comes with a wordnet lemmatizer which lammatize each word of the corpos based on its pos tags. So both verb as well as noun form are considered for lemmatization.

- cars ⇒ car ⇒ lemmatized (noun)
- bedrooms ⇒ bedrooms ⇒ lemmatized (noun)
- is ⇒ be ⇒ lemmatized (verb)
- telling ⇒ tell ⇒ stemmed

10.2 Resources Used

Networkx: Networkx module [14] has been used for page ranking and set the max iteration parameter as 200 in power method eigenvalue solver. Networkx helps create data structures for networks. A Python class termed as 'Graph' has been used for this purpose.

Table 1 Dataset description

Folder	Content
Topics	It contains files with the data extension. These files contain a topic and the filename actually describes the topic. The files have a set of sentences relevant to the topic given
Summaries-gold	This consists of summaries created by humans summaries) which are termed as gold standard. There are 4 summaries approximately for each topic

word2vec & glove & paragraph vectors: The page rank requires a weight value between two nodes, which signifies similarity between the nodes representing sentences.

$$sim(s_i, s_j) = correlation(vec(s_i), vec(s_j))$$
$$+ \text{overlapped\_tokens}(\text{lemmatized\_stemmed}(s_i), lemmatized\_stemmed(s_j))$$
$$+ \text{overlapped\_postags}(s_i, s_j)$$

$$(12)$$

Here $sim(s_i, s_j)$ denotes the similarity index of sentence i and j. The similarity value is used as weight between the nodes representing sentence(i) and sentence(j). Since page rank module was not working for negative weights, and correlation can have value in range -1 to 1, hence the correlation has been normalized to 0–1. All other parameters like overlapped number of tokens between the lemmatized and stemmed version of both the sentence i and j are also consider.

Since same pos tags of sentences can possess some similarity, hence it has also been used it as a measure for computing similarity between sentences.

Now since some sentence may not have many overlapped tokens still they may be similar, or in some cases in spite of token overlap they me be different considering negation case. Correlation has been used as a similarity measure between the vector representation of sentences.

For sentence vector generation both word2 vec and doc2vec are used as two different modules.

For constructing sentence vector the algorithms used are already discussed in earlier Sect. 7.1.

As mentioned in the algorithm, estimated frequency p(w) of every words have been found from datasets (enwiki, poliblogs, commoncrawl, text8) [1]. The parameter "a" for our task is fixed at $3 * 10^{-3}$. The vector of all other words that are not present in the collection are randomly initialized. The performance can be tuned by using different values of a. Our very first model uses this algorithm as discussed above for sentence generation.

Paragraph vectors have also been used Sect. 7.1.1 for sentence vector generation. A trained doc2vec model [17] has been used for this purpose. Two corpus have been used for training purposes. First is WIKI, which is the complete collection of English Wikipedia and second is AP-NEWS, which is a collection of news articles from the year 2009 to 2015.

Gensim: For loading both trained google news word2vec and trained paragraph vector gensim [29] is used. Gensim provides methods to load word vector or doc2vec vectors saved in binary or txt format.

After convergence of page rank algorithm, it gives every sentence a score. After that all the sentences are sorted reversely according to their ranked score. The top 7–10 sentences are selected and maintained the original order as present in the document.

Model description: Taking into account of above concepts elaborated earlier and in the section above three models have been implemented.

Table 2 Rogue—2 scores

Model No.	Description	Rogue—2 score (%)
1	Glove + word overlap + pos tagging	24
2	Word2vec + word overlapping + pos tagging	25.9
3	Paragraph + word overlap + pos tagging	26.7

Model 1: The very first model uses the inbuilt Glove[7] vectors train on wikipedia2014 and Gigaword corpus, for sentence vector generation. These word vectors are weighted averaged using the estimate word frequency as discussed above, then sentence vectors are generated under smoothing invariance factor scheme in Sect. 7.1 (jurisdiction needed).

Model 2: Model is similar to model 1, instead of using Glove vectors Google news Word2vec has been used. Also for calculating similarity between sentences pos tags overlap between sentences are considered (jurisdiction needed).

Model 3: Model 3 is different from Model 1 and Model 2 in a sense that for sentence generation it does not consider weighted average of word vectors, rather it generates word vectors using paragraph vector in Sect. 7.1.1. So in Model 3 similarity scores between sentences is calculated from the normalized values of correlation similarity between the sentence vectors, word overlap and pos tags overlap.

10.3 Evaluation

Our model is evaluated using the following process. Initially all the human generated summary of each document were merged. Thereafter the Rouge score of our summery against the merged summary was calculated. The evaluation measure rouge [19] can be defined by the following co-occurrence statistics:

$$ROUGE - N = \frac{S \in ReferenceSummaries \, Count_{match}(gram_n)}{S \in ReferenceSummaries \, Count(gram_n)}$$

n is the length of the $n - gram$, gram, and $Count_{match}(gram_n)$ is the maximum number of $n - grams$ common in a system summary and a reference summary (Table 2).

---

[7]https://nlp.stanford.edu/projects/glove/.

11 Abstractive Summary Generation

11.1 Dataset Description

The data used for this work is CNN and Daily mail news dataset. This dataset contains more that 300,000 document.

Each document contains news article and corresponding 3–5 news headlines.

583,474 news and headline pairs were extracted. From this corpus every punctuation is removed, every number is replaced with "num" token and all words are converted to lower case.

11.2 Sentence Length Selection

The minimum and maximum length of both article and headline had to be found out so that a trade-off can be achieved between number of training samples and length of samples. A carefully selected sentence length for every sentence preserves less number of padded sequence, so model becomes less bias towards padded sequences. In the current work, the minimum and maximum length of the first sentence of the article is fixed as 13 and 41 respectively. The minimum and maximum length of the length of the headlines are chosen as 7 and 16 respectively.

It was assumed from earlier sections that a constant length both for the inputs have to be specified that needs to be fed to encoder and to the labels.

So maximum length for both input (article) and label (headline) is selected as 41 and 16 respectively. Every sentence that belongs to article are either padded if its length is length is less than 41, or truncated if its length is greater than 41.

Similarly every sentences that belongs to headlines are either padded or truncated. After doing so almost 430,631 article and headline pairs were obtained.

11.3 Vocabulary Formation

Unknown words in the vocabulary should not exceed 5%. As discussed earlier, there should be a trade-off between sequence length and token used for padding. 118,649 unique words were extracted from the dataset. Every word which occurred only once, were removed from the vocabulary giving vocabulary size 86,849 words. Some parameters like Eos (end of sentence), bos (beginning of sentence) and unknown tokens were also included. A special token $< num >$ is used for all tokens which were actually numbers. A mapping dictionary between word to its index is also maintained from which every word in the dataset has been replaced with its corresponding index.

12 Analysis and Observation

Two mechanisms have been presented, one for extracting salient sentences from a document as it is, known as extractive summarization, and other one is representing the semantic information of a long sentence in shorter one by a trained model.

Analyzing extractive summarization:: Lets first talk about the model that is discussed in Sect. 10. Both the model used here is extracting salient sentences from a document. The major thing to notice here is these models actually not only consider the syntactic or lexical similarity between sentences, but these are also taking account of both structural and semantic similarity between sentences. One more thing to notice here is that these models gave improved rogue score if token overlap is considered on the lemmatized or stemmed token between sentences. The value of "a" Sect. 7.1 is also responsible for generating quality sentence vector. So the model performance can also be tested on different values of "a". Other methods that can also be used for extractive summarization could be clustering, in it we can actually make clusters of sentence vectors generated by either of the two model, and apply page rank in all the formed cluster, and take few sentences from each of the cluster. It can be noticed from the result that the paragraph vectors performed pretty well. The advantages of using paragraph vectors could be that they can be trained using data which does not have a label. They also take into account the semantics of all the words.

Analyzing Abstractive summarization:: Now lets talk about the model used for abstractive summarization. The sequence to sequence architecture is a complex architecture, which is actually trained on the first sentence of article and headline of the article. The underlying hypothesis is to generate a shorter sentences representing almost all semantic information for a given longer input sentence. There are a number of things to notice in this model. The length of the input sentence (articles first sentence) and output (headline) is selected so that the dataset does not have high number of pad symbols and sentence should not lose important information in case of truncation. The number of unknown token in our vocabulary should not be too high, otherwise model will be biased towards the unknown token. Since the vocabulary size of our model was too high hence we used sampled softmax to calculate the loss so that to increase the training speed. The rogue score computed on our test set was not pretty good. By analyzing the output computed on test set it is noticed that the model has produced at lot of pad and unknown token. To remove unknown words, they could have been replaced with the word in the input sentence which was actually getting more attention at time while predicting this word. To prevent this problem beam search could be used taking a beam size of k words at each time step of output as discussed in Sect. 8.7.

Table 3 Seq2seq model parameters

Parameters	Description	Value
seq_length_x	Length of the sequence at encoder side, i.e. is maximum number of time stamps at encoder	41
seq_length_y	Length of the sequence at decoder side, i.e. maximum number of time stamps at decoder	16
Vocabulary size	Unique (Total number of words for first sentence of article + Total number of words for headlines + num + unk + pad + eos + bos)	86,854
Embedding dimension	Dimension of vector of each words	512
Depth	Numbers of layers at encoder as well as decoder side	3
Memory dimension	Dimension of the hidden vector of all LSTM cell	512
Keep_prob	Parameter used for dropout only at training time	0.5
Number of samples	Used for computing sampled softmax loss	5
Learning rate	The weight values are updated by updating step sequences	0.001
Optimizer	Algorithm to be used to minimize the loss	Adam
Loss function	Sequence loss—average log-perplexity per symbol	–

13   Training Details

Tensorflow TensorFlow is used for implementing machine learning functions using graphs. In the graphs the nodes are the mathematical operations and the edges are data arrays (tensors) which are communicated between the respected nodes.

Table 3 mentions all the hyperparameter values used for training. Maximum length of input sequence is kept as 41 (i.e 41 words including eos[8] token), similarly length of output sequence is kept as 16 (including eos token). The number of unique words in the vocabulary set was 86,850. The vocabulary only included those words which occurred more than once in the corpus. Four extra tokens pas, unk,[9] eos, bos[10] are also added to the vocabulary. The dropout probability is kept as 0.5 while training, whereas it is kept as 1 while predicting, this hyperparameter prevents model from overfitting. Similarly sampled softmax is used at the time of training, for evaluation simple softmax is used.

14   Conclusion

Most of the work done in single document summarization is Extractive in nature. The ranking methodology described in this work for extracting important sentences in a document, requires the use of a good quality similarity index measure. This measure

---

[8]End of sentence token.

[9]Unknown token.

[10]Beginning of sentence.

depends on the accumulated result of syntactic similarity and semantic similarity between sentences. It can be seen that, by capturing semantic similarity using word vectors and paragraph vectors, the results can actually be improved. The underlying assumption of our work was to generate abstractive summary by capturing good quality extractive summary. A deep neural network model for generating abstractive summary was trained for every sentence present in the corresponding extractive summary. The quality of data set was not that fine, and there wasn't enough data sample for the model to generalize, hence it is not performing pretty well, it is giving lot of unknown and pad symbol as output word. It was noticed from the loss function that model loss was still decreasing, hence the model required more time to train. Further careful selection of training sample could have solved our problem a little bit.

References

1. Arora S, Liang Y, Ma T (2017) A simple but tough-to-beat baseline for sentence embeddings. In: International conference on learning representations. To Appear
2. Bahdanau D, Cho K, Bengio Y (2014) Neural machine translation by jointly learning to align and translate. arXiv:1409.0473
3. Barzilay R, McKeown KR, Elhadad M (1999) Information fusion in the context of multi-document summarization. In: Proceedings of the 37th annual meeting of the association for computational linguistics on computational linguistics. Association for Computational Linguistics, pp 550–557
4. Burges C, Shaked T, Renshaw E, Lazier A, Deeds M, Hamilton N, Hullender G (2005) Learning to rank using gradient descent. In: Proceedings of the 22nd international conference on machine learning. ACM, New York, pp 89–96
5. Clarke J, Lapata M (2008) Global inference for sentence compression: an integer linear programming approach. J Artif Intell Res 31:399–429
6. Daumé III H, Marcu D (2006) Bayesian query-focused summarization. In: Proceedings of the 21st international conference on computational linguistics and the 44th annual meeting of the association for computational linguistics. Association for Computational Linguistics, pp 305–312
7. Edmundson HP (1969) New methods in automatic extracting. J ACM (JACM) 16(2):264–285
8. Erkan G, Radev DR (2004) Lexrank: graph-based lexical centrality as salience in text summarization. J Artif Intell Res 22:457–479
9. Fung P, Ngai G (2006) One story, one flow: hidden Markov story models for multilingual multidocument summarization. ACM Trans Speech Lang Process (TSLP) 3(2):1–16
10. Ganesan K, Zhai C, Han J (2010) Opinosis: a graph-based approach to abstractive summarization of highly redundant opinions. In: Proceedings of the 23rd international conference on computational linguistics. Association for Computational Linguistics, pp 340–348
11. Genest PE, Lapalme G (2011) Framework for abstractive summarization using text-to-text generation. In: Proceedings of the workshop on monolingual text-to-text generation. Association for Computational Linguistics, pp 64–73
12. Genest PE, Lapalme G (2012) Fully abstractive approach to guided summarization. In: Proceedings of the 50th annual meeting of the association for computational linguistics: short papers, vol 2. Association for Computational Linguistics, pp 354–358
13. Greenbacker CF (2011) Towards a framework for abstractive summarization of multimodal documents. In: Proceedings of the ACL 2011 student session. Association for Computational Linguistics, pp 75–80

14. Hagberg AA, Schult DA, Swart PJ (2008) Exploring network structure, dynamics, and function using NetworkX. In: Proceedings of the 7th python in science conference (SciPy2008). Pasadena, CA USA, pp 11–15
15. Haghighi A, Vanderwende L (2009) Exploring content models for multi-document summarization. In: Proceedings of human language technologies: the 2009 annual conference of the North American chapter of the association for computational linguistics. Association for Computational Linguistics, pp 362–370
16. Kupiec J, Pedersen J, Chen F (1995) A trainable document summarizer. In: Proceedings of the 18th annual international ACM SIGIR conference on research and development in information retrieval. ACM, New York, pp 68–73
17. Lau JH, Baldwin T (2016) An empirical evaluation of doc2vec with practical insights into document embedding generation. arXiv:1607.05368
18. Le Q, Mikolov T (2014) Distributed representations of sentences and documents. In: Proceedings of the 31st international conference on machine learning (ICML-14), pp 1188–1196
19. Lin CY (2004) Rouge: a package for automatic evaluation of summaries. In: Proceedings of the ACL-04 workshop on text summarization branches out, vol 8. Barcelona, Spain
20. Lin CY, Hovy E (2000) The automated acquisition of topic signatures for text summarization. In: Proceedings of the 18th conference on computational linguistics, vol 1. Association for Computational Linguistics, pp 495–501
21. Liu F, Flanigan J, Thomson S, Sadeh N, Smith NA (2015) Toward abstractive summarization using semantic representations
22. Liu F, Liu Y (2009) From extractive to abstractive meeting summaries: can it be done by sentence compression? In: Proceedings of the ACL-IJCNLP 2009 conference short papers. Association for Computational Linguistics, pp 261–264
23. Luong MT, Sutskever I, Le QV, Vinyals O, Zaremba W (2014) Addressing the rare word problem in neural machine translation. arXiv:1410.8206
24. Moawad IF, Aref M (2012) Semantic graph reduction approach for abstractive text summarization. In: 2012 seventh international conference on computer engineering & systems (ICCES). IEEE, pp 132–138
25. Nallapati R, Zhou B, Gulcehre C, Xiang B et al (2016) Abstractive text summarization using sequence-to-sequence rnns and beyond. arXiv:1602.06023
26. Page L, Brin S, Motwani R, Winograd T (1999) The pagerank citation ranking: bringing order to the web. Technical report, Stanford InfoLab
27. Qazvinian V, Radev DR (2008) Scientific paper summarization using citation summary networks. In: Proceedings of the 22nd international conference on computational linguistics, vol 1. Association for Computational Linguistics, pp 689–696
28. Radev DR, Jing H, Styś M, Tam D (2004) Centroid-based summarization of multiple documents. Inf Process Manag 40(6):919–938
29. Řehůřek R, Sojka P (2010) Software framework for topic modelling with large corpora. In: Proceedings of the LREC 2010 workshop on new challenges for NLP frameworks. ELRA, Valletta, Malta, pp 45–50. http://is.muni.cz/publication/884893/en
30. Rush AM, Chopra S, Weston J (2015) A neural attention model for abstractive sentence summarization. arXiv:1509.00685
31. Sutskever I, Vinyals O, Le QV (2014) Sequence to sequence learning with neural networks. In: Advances in neural information processing systems, pp 3104–3112
32. Svore KM, Vanderwende L, Burges CJ (2007) Enhancing single-document summarization by combining ranknet and third-party sources. In: Emnlp-conll, pp 448–457
33. Vinyals O, Fortunato M, Jaitly N (2015) Pointer networks. In: Advances in neural information processing systems, pp 2692–2700
34. Wiseman S, Rush AM (2016) Sequence-to-sequence learning as beam-search optimization. arXiv:1606.02960

Semantic Web and Data Visualization

Abhinav Singh, Utsha Sinha and Deepak Kumar Sharma

Abstract With the terrific growth of data volume and data being produced every second on millions of devices across the globe, there is a desperate need to manage the unstructured data available on web pages efficiently. Semantic Web or also known as Web of Trust structures the scattered data on the Internet according to the needs of the user. It is an extension of the World Wide Web (WWW) which focuses on manipulating web data on behalf of Humans. Due to the ability of the Semantic Web to integrate data from disparate sources and hence makes it more user-friendly, it is an emerging trend. Tim Berners-Lee first introduced the term Semantic Web and since then it has come a long way to become a more intelligent and intuitive web. Data Visualization plays an essential role in explaining complex concepts in a universal manner through pictorial representation, and the Semantic Web helps in broadening the potential of Data Visualization and thus making it an appropriate combination. The objective of this chapter is to provide fundamental insights concerning the semantic web technologies and in addition to that it also elucidates the issues as well as the solutions regarding the semantic web. The purpose of this chapter is to highlight the semantic web architecture in detail while also comparing it with the traditional search system. It classifies the semantic web architecture into three major pillars i.e. RDF, Ontology, and XML. Moreover, it describes different semantic web tools used in the framework and technology. It attempts to illustrate different approaches of the semantic web search engines. Besides stating numerous challenges faced by the semantic web it also illustrates the solutions.

A. Singh · U. Sinha · D. K. Sharma (✉)
Department of Information Technology, Netaji Subhas University of Technology (Formerly Known as NSIT), New Delhi, India
e-mail: dk.sharma1982@yahoo.com

A. Singh
e-mail: abheesing@gmail.com

U. Sinha
e-mail: utsha.sinha1510@gmail.com

© Springer Nature Switzerland AG 2020
J. Hemanth et al. (eds.), *Data Visualization and Knowledge Engineering*,
Lecture Notes on Data Engineering and Communications Technologies 32,
https://doi.org/10.1007/978-3-030-25797-2_6

Keywords Uniform resource locator · Facebook · Artificial intelligence ·
Resource description framework schema · Web ontology · Language · Application
program interface

1 Introduction

There is huge, endless, and constantly increasing amount of data available in the
world. The current generation lives in the blurry zone which is flooded with data.
Personal computers make the affair rather efficient and help us easily save our data.
Data mining using computers helps in automating or at least simplifying the search on
data. Thus, storage mediums need to use data mining to attain better search strategies.
Syntactic mode of discovery becomes a limitation which is what the current web
service depends upon. However, this limitation leads to the development or evolution
of Web 3.0 (Semantic Web) using the conventional mode of service. As can be seen, in
order to overcome this drawback, semantic web develops [1]. *Tim Berners-Lee* who
also founded WWW (World Wide Web), URIs (Uniform Resource Locators), and
HTML (Hypertext Markup Language) coined the term Semantic Web. The Semantic
Web which is also considered an extension of the World Wide Web which utilizes
the web not only according to the needs of Human but also enables the machine to
understand it efficiently and intelligently so as to enhance the desired search result.
Through Semantic Web, people are given the chance to augment the collection of
data by adding onto the existing collection. One of the idiosyncratic features of the
Semantic Web is that in addition to humans, it can also be read by machines as well
i.e. it is machine interoperable.

Access to information on the web that is distinct to each user is required to filter out
the results provided by different search engines (e.g. Google and Yahoo). It becomes
very difficult for users to make sense of the multiple information resources, divided
into tiny fractions, available on the internet. According to the Ref. [2], as of 2008,
one trillion pages of the web's portion were already accessible by the search engines.
Users often become overwhelmed and may or may not find what they are looking
for, making searching cumbersome. It is also important to realize that the scale of
the web is highly decentralized, redundant, and imprecise in nature. The information
available, no matter how useful, is scattered over various resources, making the whole
process even more complex [2].

This chapter talks about the Semantic Web and the architecture involved. While it
focuses on the upcoming Semantic Web technologies, it also explains the limitations
of the traditional web. It offers a systematic as well as a simple approach of the
aforementioned concept for the starters who intends to pursue an interest in the
Semantic Web research.

1.1 Background

Since most of the search engines return results based on the recent activities of the users and does not index the whole database of the web, it becomes the need of the hour to have a fully functional unified search engine which is able to explore the whole World Wide Web before producing its result. There are two types of issues faced by any search engine while working with the information given on the web. Firstly, the challenge to map the query to the information available on the Internet and produce desired and intelligent results. Secondly, the obstacle to efficiently identify the preferred data from the given linked documents [3]. The semantic web is able to tackle the former problem with the help of semantic annotations. The latter issue can be solved by the models based on graph-query [4]. Semantic web technologies are becoming increasingly popular and penetrating the industrial market. Organizations like Schema.org and Facebook are manipulating the utilities of metadata and attempting to guide on the web. Semantic technologies are being followed by Apple's Siri and IBM's Watson. The most used search engine on the web, Google is attempting to structure its search engine in a more semantic manner [5]. Semantic web focuses on the meaning of the data rather than emphasising on the syntactic structure. The semantic interoperability associates the known with the unknown terms making it an efficient method to link data [6]. Various researches are being carried out on the ontology part of the semantic web since it is the backbone of the semantic web. Takahira Yamaguchi 's Group is integrating Artificial Intelligence with the semantic web to achieve full independence. They are working on developing a system capable of answering while utilizing the data and ontologies. Moreover, they were able to combine robots with the semantic web [7].

This chapter aims to highlight the semantic web architecture in detail while also comparing it with the traditional search system. Moreover, it describes different semantic web tools used in the framework and technology. Besides stating numerous challenges faced by the semantic web it also illustrates the solutions.

1.2 The Semantic Web

Meaning to a variety of web resources can be provided using Semantic Web, which allows the machine to interpret and understand the data according to the web users' requests and to their satisfaction. Semantic Web is the extended and improved version of the World Wide Web (WWW) and is referred to as WWW 2.0. The original idea of a second generation web was given by W3C and WWW's founder, *Sir Tim Berners-Lee*. According to *Jeon and Kim*, this upgraded version, using the meaning of web, provides the users with the ability to share their data beyond any limitations and barriers with regard to programs and websites [8]. Unlike the Relational Database, Semantic Web is concerned with the meaning of the data and not its structure. It is the recent revolution in the field of Information and Technology. It helps computers understand the data and link it from source to source. Semantic Web applications are different from other applications on the basis of 3 technical standards [1]:

- Semantic Web uses the Resource Description Framework as a data modelling language. It utilizes URIs and text strings to store as well as represent its information in the RDF format.
- RDF Querying Language and SPARQL are the two methods to retrieve data from the systems.
- One of the most important languages used in the Semantic Web is the Web Ontology Language also used as schema language and representation language (5).

Even though a large number of traditional search engines are syntactically accurate but the amount of results is enormous making the search tedious. Therefore, to solve the aforementioned issue Semantic search engine depicts the information in the form of accurate and unambiguous expressions. The selective construct of semantic web enables the users to find anything with ease with according to their requirements. A Considerable amount of web engines have been developed till now using the semantic web architecture as their backbone but they differ in terms of end results (17). According to the Ref. [4] the semantic search consists of four major approaches given below:

1. The **first** and foremost approach eliminates the ambiguity from search queries through contextual analysis.
2. The **second** approach uses reasoning to infer certain additional logical facts from the given set of facts.
3. The **third** approach emphasizes the understanding of natural language in which the search engine attempts to find the actual purpose of the information with the help of the user-submitted query.
4. The **fourth** approach chooses Ontology i.e. domain of knowledge as its main function to broaden the scope of the search query (Fig. 1).

1.3 Semantic Web Technologies

Considering the popularity of the semantic web, various developers are attempting to develop different semantic search engines which can be incorporated by the present search engines. Glauco Schmidt and Alcides Calsavara suggested an innovative search engine service based on semantic web. It accumulates a broad set of data available on the web in a form readable to the machine so as to enhance its efficiency and accuracy. The aforementioned feature helps the search engine to give an accurate response to a puzzling query [4]. The following examples of the semantic web search engine are based on the meaning and structure of the query, unlike the traditional search engines which largely depends on the keywords grouping and inbound link measurement algorithms.

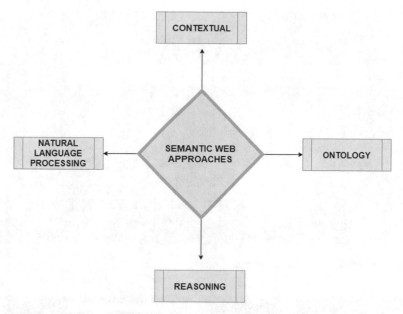

Fig. 1 Types of semantic web approaches

1. **Kosmix**: It enables the users to retrieve only relevant information about any particular field by presenting them with a table of contents called 'Your Guide to the Web'.
2. **Exalead**: It uses a unique process of filtering the results of queries by image-based features. It reduces the results of the search through the colour, size and content.
3. **SenseBot**: This type of search engine generates a summary of the desired query which contains various URLs to different web pages which might satisfy the user's need.
4. **Swoogle**: It firmly follows the guidelines of the semantic web. It stores the data containing diverse documents stored in the format of RDF.
5. **Lexxe**: It uses Natural Language Processing to operate the question-answering type of semantic web.
6. **DuckDuckGo**: It is able to search smartly throughout the web and if the requested query carries more than one meaning, it displays the results of every type which eventually helps in choosing the desired result.

All the above-mentioned examples heavily rely on a certain number of factors like relevancy of data, fast response time, accurate result production and user-friendliness (Figs. 2 and 3).

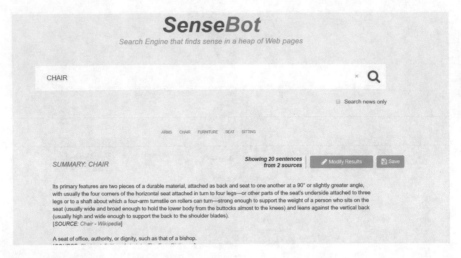

Fig. 2 Screenshot of SenseBot result page

Fig. 3 Screenshot of kosmix result page

2   Traditional Search Engine

The current World Wide Web possesses a vast amount of information yet lacks a basic logical structure. Owing to the capacity of the WWW, it works on the fundamental of keyword-based searching. Despite the fact that the World Wide Web is considered to be a huge informational database but it produces obscure and ambiguous results since the machine is not able to accurately interpret the given information. The next section describes the limitations of the World Wide Web due to which semantic web was introduced.

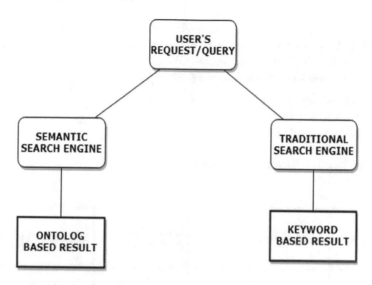

Fig. 4 Working of semantic search engine & traditional search engine

2.1 Limitations

The traditional search engine has various drawbacks which are given below:-

1. The information presented to the user's queries is deficient of logical structure.
2. Vagueness and uncertainty in results due to inefficient data linkages.
3. Inability to cater to a large number of users in a systemized manner.
4. Lack of machine understand ability factor.
5. The absence of data management.
6. Incapacity to utilize resources efficiently (Fig. 4).

2.2 Intelligent Semantic Web

The semantic web not only helps in automating the search process but also give better results than the traditional search engines based on the ontology, machine readability and efficient searching. It is able to distinctly identify the objects and the connections between them which enable it to make an efficient and faster search [9]. For example, if the requested query is about a movie theatre then the semantic search engine will produce the result faster and with more accuracy, if there is a unique identifier like show timings, location to distinctly illustrate the relation between them.

An easy and understandable model of the key aspects of the kinds of information, also known as ontologies, is essential for machines to be able to automatically learn that information. In order to efficiently execute it on a web-scale level, proper technical standards are presented which will establish the appropriate format for it. Some of the languages determining the apt format are RDF (Resource Description Framework), RDFS (RDF Schema), and OWL (Web Ontology Language) [9]. All in all, the semantic web has the following advantages:

1. Efficient linking of data.
2. Result simplification.
3. Outcome accuracy and transparency.
4. Logical structure.
5. Comprehensive resource utilization (Fig. 5).

SL NO.	COMPONENTS	TRADITIONAL SEARCH ENGINE	SEMANTIC WEB
1.	RESOURCE UTILISATION	✖	✔
2.	LOGICAL STRUCTURE	✖	✔
3.	DATA MANAGEMENT	✖	✔
4.	MACHINE UNDERSTANDABILITY	✖	✔
5.	EFFECTIVE INTERCONNECTION OF WEB INFORMATION	✖	✔

Fig. 5 Comparison of traditional search engine and semantic web

3 Semantic Web Architecture

In order to achieve a certain level of "intelligence" on the web, the semantic web was developed which processed the information with high efficiency and accuracy. The general architecture of the semantic web includes a user interface which interacts with the users and represents the information, an RDF i.e. Resource Description Framework which is a language designed to characterize the data in a way that machines are able to understand and interpret it, the OWL i.e. Web Ontology Language which is used to enhance the lexicon and the relations between distinct objects and classes. The next section describes the aforementioned components of the semantic web (Fig. 6).

Fig. 6 Architecture of semantic web

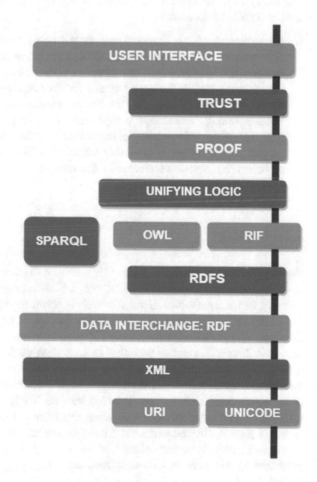

3.1   Resource Description Language

Resource Description Framework (RDF) being a W3C illustrates the web resources. RDF is not intended for human presentation but for computer reading and understanding purposes. RDF is prominently written on XML. RDF, being a simple data model language describes the relationships between different objects. It has the ability to build models in various syntaxes such as XML, N3, Turtle, and RDFa. The vocabulary part and the class description side is managed by the RDF Schema [6]. RDF delivers two essential features i.e. interchangeability and scalability which promote the semantic web development. Since the RDF data is majorly written in the XML format, it makes it scalable as well as interchangeable which eventually help it to store a large set of data with ease [10]. According the Ref. to [10] RDF is made up of three vital elements:

1. **Resources**: Any web site, web page or even object which can be represented by RDF and is uniquely identified by URI is known as resource.
2. **Properties**: The main task of properties is to explain the relationships between the resources and its features, element, trait or any distinctive characteristic.
3. **Statements**: It built upon the link between various resources and their features. It has essentially three major components: subject, predicate, and an object. The first part of the RDF which selects which resource to be described is called a subject. The predicate is used to describe the relationship between a subject and an object. An object is the value of the subject.

3.2   Ontology

Ontology Model in Semantic Web is very similar to concept mapping in the human brain, and it offers efficiency. For example, a student may refer to old topics to make meaning of the new topics. Semantic Web is about the relationship of one thing to the other and how it happened. It's not just limited to pages and links. The idea of Semantic Web came into existence in 2003, and no major advancement has yet been observed. It is important to note that factors such as dot com burst and huge data model play a role, and www 2.0 is a growing research field [1]. There are many ontology-based search systems such as which was described by Georges Gardarin et al. which converted the web sources into an XML construction. The required information is fetched through Web Wrappers from the vast options available on the web and then finally the extracted data is processed using data mining techniques like classification. Another example of an ontology-based search engine is the one discussed by Maedche et al. which integrated searching, updating, and processing. It consists of two methods by which the ontology searching is performed-(1) query-by-example and (2) query-by-term [4].

3.2.1 Oil

OIL or Ontology Inference Layer was developed with an aim to integrate ontologies with the current web system and is also known as the extension of RDF. It depends on three major factors: Frame Based System, Description Logics, and XML. The main purpose of the frame based system is to provide modelling primitives, for instance, subclass-Of whereas the semantics are inherited from the description logics. XML provides the syntax through DTD and XML Schema. While OIL can be written in text files it can also be understood by RDF processors, partially [10].

3.2.2 Daml + Oil

DARPA project DAML resulted in the development of DAML + OIL (DARPA Agent Markup Language + OIL) which is an ontology language initiated by a joint committee of the USA and the European Union. A diverse range of people took part in this project including software developers, government organizations, researchers, and the W3C groups. The project focused on providing an ontology language to the semantic web. DAML + OIL before reaching its final version went through the various version and the earliest of the version was known as DAML-ONT. The DAML + OIL was based on the object-oriented technique which categorized classes on the basis of their domains. It also shared many common properties of OIL [10].

3.2.3 Owl-S

Web ontology language for services (OWL-S) or formerly known as DARPA Agent Markup Language for Services (DAML-S) is used to describe the characteristics and abilities of web services. It is a W3C standard submitted in 2004. According to the Ref. [1] it constituents of the following three features:

1. **Service Profile**: It links the user with the provider as the user defines his needs and the provider enumerates his services.
2. **Service Model**: It is responsible for the smooth execution, production and composition of the process.
3. **Service Grounding**: It synchronizes the usage of services.

3.3 Extensible Mark-up Language

It is a scripting language used to store, structure and transfer data in the form of tags. It enables the user to design its own tag which is called a root tag and it can be used anywhere throughout the database. It is used in various applications such as Android development, file generator. Its flexible structure enables it to send as well as receive the data from the server. The ability to define its own tags helps the semantic search engine to easily identify the objects and relations between them.

XML is being widely used for internet interoperability. It provides the basis for combining multiple elements into a specific function from a different vocabulary [6].

3.4 Logic, Trust and Proof

The main function of the logic layer is to further augment the ontology language and to enable application-specific declarative knowledge to be written. The proof layer plays a major role in the authentic deduction of the data while it also represents the proofs in web languages. The trust layer ensures the data is certified from trusted sources such as digital signatures or certified agencies [11].

3.5 Applications

According to the Ref. [8] there are broadly 5 applications of the semantic web which are described as follows:

1. **Data Management**: A huge amount of knowledge is scattered throughout the web and semantic web is capable of restructuring it efficiently.
2. **Business Development**: Experts in the e-commerce industry believe that the semantic web is convenient and fully functional in terms of e-business match-making i.e. a process of uniting the business with its potential client.
3. **Automation**: Web services enabled by semantic web have the ability to optimize and automate tasks ranging from a simple task to complex task. For example, searching and booking of a hotel become an effortless task through the semantic web and web services.
4. **Knowledge Discovery**: The semantic web can be used to develop a separate database for a company consisting of the information about the past, present and expected projects of the company which will not only help the employees to search for suitable projects but also helps the organization to tack them.
5. **Decision Support**: Work is currently underway to develop semantically—enabled decision support systems (DSSs) that help the end user to make the decision more efficiently (Fig. 7).

4 Data Mining

Since web mining combines the two most important aspects of web i.e. Data Mining and World Wide Web, it is naturally a booming research topic. The World Wide

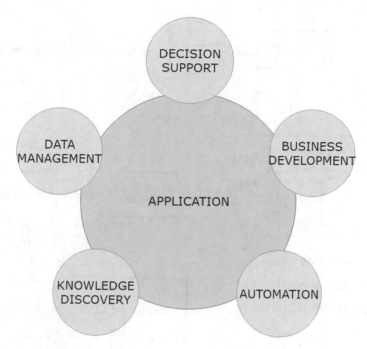

Fig. 7 Applications of semantic web

Web is an immense source of information which makes it an ideal match for the data mining industry. The research related to web mining involves various research communities, like database and AI. Even though the web is enormous, diverse and dynamic but it also poses certain issues like temporal problems and high multimedia data. Etzioni introduced the term web mining at first in his research paper which was published in 1996. The next describes web mining in the context of the semantic web [2].

4.1 Semantic Web Mining

As the name suggests it is a combination of two scientific areas: (1) Semantic Web (2) Data Mining. While the semantic web is implemented to offer to the data making a complex heterogeneous data structure whereas data mining is used to extricate patterns from the homogeneous data. Diverse knowledge areas like Bio-medical can be transformed into a perfect target to be mined leading to the evolution of semantic web mining owing to the fact that there is an increase in the stored semantic data. To get more insights and explore more possibilities in semantic annotations it's a good idea to mine semantic web ontologies. The huge amount of semantic data currently stored in a complex way requires a more sophisticated and efficient way to be stored

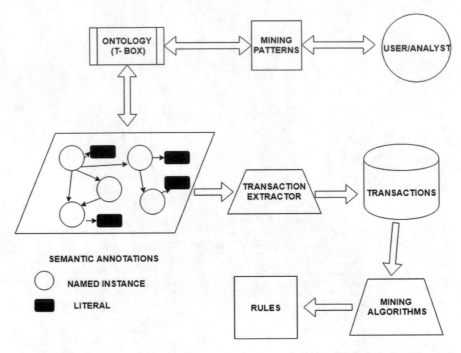

Fig. 8 Data mining of semantic web [9]

or to extract them which is provided by the data mining techniques. For example, Twine's semantic web portal service stores the users' data using OWL and RDF which makes up for more than three millions of semantic tags [8]. Figure 8 shows the working of data mining through semantic web.

4.2 The Drawbacks of Semantic Web Mining

Even though semantic web mining is proved to be a useful technique to mine the semantic data but it poses some challenges because of the complexity and hetero-geneity of the semantic data. According to the Ref. [9], the leading most barrier in mining semantic web is transaction recognition and identifying semi-structured data due to the following three reasons: (1) Traditional data mining algorithms are based on homogeneous data but on the contrary the semantic web contains heterogenic data. (2) The semantic data is represented using a subject, predicate, an object which is a three-layered architecture, making the data complex. (3) The description logics in OWL sublanguages makes the data more complex and heterogeneous. Reference [9] stated that the structure of semantic web ontology poses a challenge when the traditional tree algorithm is applied to it or when the mining algorithm attempts to stipulate variables but the ontology creates a scenario of an unlimited number of properties which then goes on to create multiple values. Due to the reduction in the threshold of the mining process caused by the non-frequent items can lead to the production of irrelevant patterns.

5 Semantic Web Tools

Various components are used in constructing semantic web which can be classified as the ontology editors, semantic web reasoners, and the semantic web APIs. Each of the tools is described in the next section of the chapter.

5.1 The Ontology Editor

Applications configured to produce or operate ontologies are known as the ontology editors. Some of the examples of the ontology editors are Protégé, NeOn Toolkit, SWOOP, and Fluent Editor. In a collaborative project with the University of Manchester, Protege was produced at Stanford University. The free open-source platform not only lets the user develop domain models but also directs them to create applications through ontologies. The core principle which the protege is based on is the massive amount of modelling structure it possesses. Due to its user-friendly features, it is considered to be the most used ontology editor [10].

5.2 The Semantic Reasoner

It is a software system that extracts the analytical and rational component from the given information. The rule with which the semantic reasoner analyses the content is constructed on the ontology language. Few of the popular examples of such reasoners are CB, CEL, FaCT ++, Hermit, Pellet, RP, SR, TrOWL. The Fig. 9 illustrates the comparison between the aforementioned semantic reasoners [10].

5.3 The Semantic Web APIs

One way of representing the semantic web ontologies is through the OWL API which is a Java interface for OWL. According to the Ref. [10] the OWL API has the following components:

- RDF/XML Parser
- OWL/XML writer

	CB	CEL	FaCT++	HermiT	Pellet	RP	SR	TrOWL
OWL API	-	+	+	+	+	+	+	+
OWLink API	-	+	+	+	+	+	-	-
Protege PLUGIN	-	+	+	+	+	-	+	+
License	DuLi: GLGPL	AP 2.0	GLGPL	GLGPL	DuLi: AGPL	own	own	DuLi: AGPL
Open source	+	+	+	+	+	-	-	-
Language	OCaml	COMMON LISP	C++	Java	Java	Lisp	Java	Java
Platforms	all	Linux	all	all	all	all	all	all
Jena	-	-	-	-	+	-	-	-
Institution	a	a	a	a	c	c	g	a

Fig. 9 Comparison of different semantic reasoner [10]

- OWL Functional Syntax Parser
- OWL 2 API

The Jena API, being a free open source platform to construct applications based on the semantic web. One of the most important characteristics of the Jena API is that it contains libraries on an extensive level [10].

6 Further Challenges

Even though the semantic web is able to overcome the challenges of the traditional web efficiently but there remain a few common challenges that need to be addressed [6, 4].

1. **Imprecise queries**: All the potential keywords relating to the result are not incorporated by the user since they possess only domain-specific knowledge.
2. **Low precision**: Not all the semantic search engines are able to showcase accurate interpretation of the queries. For example, Ding's semantic search engine relies on the top 50 search result of Google which is in no case a semantic structure.
3. **The accurate discovery of the user's intention**: The most important factor for the semantic web is to correctly determine the intention of the user's request.
4. **Ontology Development**: Since the semantic web is largely dependent on the ontology, it creates a necessity to manage, utilize, and enhance the ontology database.

5. **The authenticity of the result**: The semantic web comprises of a vast amount of information and it relies on it to produce the request result but there is a major flaw regarding the authenticity and originality of the information.
6. **Limited content**: The information available on the World Wide Web is enormous whereas the data is limited in case of the semantic web when compared to the Internet.
7. **Scalability**: Although it is an essential task to store, produce, process, and structure the semantic web data but it becomes critical to incorporate the changes in a scalable manner in order to broaden the scope and increase the sustainability of the semantic web.

In order to develop a semantic search engine, future researchers must focus on the above-mentioned issues to increase the longevity and sustainability of the search engine.

7 Solution and Conclusion

To make the semantic web a sustainable concept, it should be integrated with the existing World Wide Web which will not only improve the scalability factor significantly but will also be able to achieve its full potential. Various challenges have been stated in the previous section of the chapter in the context of semantic web mining but to tackle that there are varied solutions. For example, in order to tackle the hidden knowledge residing in the semantic data, the semantic reasoners have been used. To solve the triple-layered architecture problem, a process is followed which at first computes the value composition and then through grouping, it constructs transactions refined according to the user's demands. To overcome issues like the unlimited propertied with unlimited values during data mining, a number of solutions have been proposed such as specifying the relations between the object's concepts and roles in the mining process. Since the non-frequent item produces an irreverent pattern, it can be solved by the generalization of item values on the basis of concepts.

Semantic web proposes a novel yet efficient way of searching and indexing the existing web and producing accurate results through the help of machines interpretations. Even though the semantic web is an upcoming field and benefits the World Wide Web in a lot of ways such as giving a logical structure to the data, simplifying the result, efficiency in linking the data, utilizing the resources in a thorough and an efficient manner, and producing an accurate yet transparent end result but it poses certain challenges which have been explained thoroughly in the previous sections of the chapter. Web mining is another aspect through which the importance of the semantic web is highlighted. Combination of web mining and semantic web creates an upcoming field which is not yet fully explored and has numerous hidden components in it. All in all, there exists many semantic search engines such as DuckDuckGo or Swoogle but there is an absence of proper efficient system utilizing the semantic web.

This chapter deals with the concept of the semantic web and at the same time presents a comparison between the traditional search engine and the semantic search engine. In addition, it also explains the architecture of the semantic web while explaining the tools of the semantic web too. Moreover, it attempts to incorporate the importance of web mining in the semantic web in order to produce a more efficient end product.

References

1. Aroma J, Kurian M (2012) A survey on need for semantic web. Int J Sci Res Publ 2(11). Coimbatore, India
2. Gopalachari MV, Sammulal P (2013) A survey on semantic web and knowledge processing. Int J Innov Res Comput Commun Eng 1(2). Hyderabad, India
3. Madhu G, Govardhan A, Rajinikanth TV (2011) Intelligent semantic web search engine. Int J Web Semant Technol 2(1). Hyderabad, India
4. Chitre N (2016) Semantic web search engine. Int J Adv Res Comput Sci Manag Stud 4(7). Pune, India
5. Bakshi R, Vijhani (2015) A semantic web-an extensive literature review. Int J Mod Trends Eng Res 2(8). Mumbai, India
6. Pandey G (2012) The semantic web. Int J Eng Res Appl 2(1). Gujarat, India
7. Bukhari SN, Mir2 JA, Ahmad U (2017) Study and review of recent trends in semantic web. Int J Adv Res Comput Sci Softw Eng 7(6). Jammu and Kashmir, India
8. Quboa QK, Saraee M (2013) A state of the art survey on semantic web mining. Intell Inf Manag 5(1). Salford, UK
9. Shabajee P (2006) Informed consent on the semantic web – issues for interaction and interface designer. In: Proceedings of the third international semantic web user interaction workshop, Athens, GA
10. Ishkey H, Harb HM, Farahat H (2014) A comprehensive semantic web survey. Al-Azhar Univ Eng J 9(1). Cairo, Egypt
11. Kathirvelu P Semantic web technology, layered architecture, RDF and OWL representation

Challenges and Responses Towards Sustainable Future Through Machine Learning and Deep Learning

Saksham Gulati and Shilpi Sharma

Abstract Over the years, sustainable development has emerged as an extensively used theory. Radically speaking, the concept of sustainable development is a bid to blend inflating concerns regarding a variety of environmental issues along with socio-economic issues. Shifts in the demeanour of individuals, institutes and firms are presumed as a prerequisite for sustainable development. Through the mode of sustainable development, we can identify our position in the ecosystem by evaluating the comprehensive influence of our actions. Sustainable development has the possibility to sermon fundamental provocations for humanity, at present and for the future. Nevertheless to achieve this, more certainty of meaning is needed. However instead of only considering it as an environmental concern, focus has also been shifted to assimilate economic and social capacities. Artificial Intelligence and it's subsets like machine learning and deep learning are highly effective in overcoming the hurdles which are blocking the path to sustainable future. This paper introduces a methodology which can be adopted to understand the goals of sustainable development by means of knowledge engineering, machine learning and overcome the hurdles for sustainable future through deep learning. Furthermore this paper is aimed at overcoming the hurdles of sustainable development by means of supervised learning techniques.

Keywords Sustainable development · Socio-economic issues · Environmental issues · Artificial intelligence · Knowledge engineering · Machine learning · Deep learning · Supervised learning

S. Gulati (✉) · S. Sharma (✉)
ASET, Amity University, Noida, Uttar Pradesh, India
e-mail: saksham.gulati28@gmail.com

S. Sharma
e-mail: ssharma22@amity.edu

© Springer Nature Switzerland AG 2020
J. Hemanth et al. (eds.), *Data Visualization and Knowledge Engineering*,
Lecture Notes on Data Engineering and Communications Technologies 32,
https://doi.org/10.1007/978-3-030-25797-2_7

1 Introduction

Environmental issues is one of the major hurdle in the path of reaching the goals of sustainable development. Air quality is one of the major concerns of poor environment and this quality of air is getting worse with the generation of high amounts of waste which goes unmanaged. This paper introduces a model which can be put in place to predict air quality on the basis of NO_2 and PM10 and another theory for better waste management through AI and automation without affecting the human health. Since waste management through human resources could be a major health hazard for current as well as future generations.

This paper has been divided into two practical tests one is done by machine learning for air pollution while other is done by deep learning. In our first scenario we have categorised the states of India which has the worst quality of air on the basis NO_2 and PM10 concluded the use of data analytics, data engineering and data visualisation. Then we have predicted the air quality index of the basis of presence of NO_2 and PM10 in the air for different years. This model can be reused to predict the air quality in upcoming time.

Sustainable development refers to the development of economic growth without depleting the natural resources [1]. The past few years have seen an exponential increase in the realisation of importance of sustainable development. Although being a critical topic of discussion sustainable development still has gaps and loopholes to completely up-ended to its true and authentic meaning. Sustainable development thus requires the participation of diverse stakeholders and prospects. They concluded a new synthesis and coordination of conjoint action in order to achieve multiple values synergistically and instantaneously. Synergistically agreeing upon the sustainable values, actions, goals and deliverables is as of now, far sighted. United Nations describes 17 goals to evaluate sustainable development including zero poverty; no hunger; Good Health; High Quality Education; Gender Equality; Clean Water and Sanitation; Affordable and Clean Energy; Decent Work and Economic Growth; Industry, Innovation and Infrastructure; Reduced Inequalities; Sustainable Cities and Communities; Responsible consumption and production; Climate Action; Life Below Water; Life on land; Peace, Justice and Strong Institution; Partnership for the goals [2].

Thus, the concept of sustainability has been adapted to address different challenges extending from planning sustainable smart cities to sustainable livings and also to sustainable husbandry. Sufficient efforts have been made to develop common corporate standards in World Business Council and UN Global Compact for Sustainability [3]. Every human being has the right to learn at each stage of their lives. Also, they continue to gain knowledge for life at work place, during travel or social relationship. Reference [4] presents a newly hypothesized learning framework that is functional to machine learning and is centred towards the characteristics of human self-governing learning.

With the increasing size and complexity of networks, efficient algorithms for deep networks in a distributed/parallel environment is presently the most impenetrable problem in academia and industry. Refer to a case study, Google used 1-billion parameter neural network and took 3 days to train over a 1000-node clusters with more than 16,000 CPU cores [5]. Each instances have 170 servers in a network. The parameters of giant deep networks in a distributed computing environment splits across multiple nodes. It results in costly communication and synchronization to transfer the parameter updates between the server and the processing nodes.

The current challenges for deep learning illustrate a great demand for algorithms that reduce the amount of computation and energy usage. To reduce the bottleneck matrix multiplications, there has been a flurry of works around reducing the amount of computations associated with them [6].

In this paper a retraining neural network has been used which has an ability to keep a checkpoint on previously trained dataset. A new dataset can be trained over the older dataset thereby retaining the ability to classify previously trained images as well as classifying new labels through a knowledge transfer process. The primary reason for using a retrain Inception Net was to reduce economical cost related to neural networks and introducing a method with better resource management with lesser waste production.

Supervised Learning is a set of machine learning and deep learning algorithms that are used to perform prediction and classification tasks by learning from the previous data. In supervised learning, a training data is classified into labels or regression data is fit into range by the respective supervised learning algorithm and then the new data is tested on the basis of such labels and a range is defined to create an accurate predictions by feeding new data to the trained model. There are usually a few types of errors which are used to find accuracy of the predicted data including mean absolute error, mean square error, R-Squared error etc.

The proposed paper is organized into different section in which Sect. 2 describes the literature review and related work, Sect. 3 represents the proposed methodology and contribution towards the paper, Sect. 4 presents the simulated experimental results, Sect. 5 states the conclusion and future scope.

2 Literature Review

With rising concerns for development of sustainable future, individuals and organisations are coming up with innovative and creative approaches. Artificial intelligence has become a recent key player in realising and reaching the goals of sustainable future. One such use case can be found in "Artificial societies for integrated and sustainable development of metropolitan systems" which has tried to solve the modern traffic problems using intelligent infrastructure. Machine Learning is a field of computer science and a subset of artificial intelligence in which a computer gains a potential to learn and improve from experience without being explicitly programmed [7]. Machine Learning uses statistical models to make a prediction based on previously fed data. Data Engineering and knowledge engineering plays a key role in implementing a predictive analysis based on machine learning. Knowledge is a concept which stands for accumulation of sensible information gathered over a period of time. One of the major advantages of knowledge engineering is that it can be shared among similar networks to expand the reachability of a single network to a larger group through a process called, knowledge acquisition quality of related networks This knowledge acquisition quality has proved to be highly economical [8, 9].

Reference [10] concludes a number of data pre-processing techniques for (a) solving data problems (b) understanding the nature of the data (c) performing in-depth data analysis. The data pre-processing methods are divided into following techniques namely Data Cleaning, Data Transformation and Data reduction. Hence it an important issue while pre-processing was considered for data analysis. Most important problems in data pre-processing is to know what valuable information exists in the raw data and where to preserve. Involvement of the domain expert results in some useful feedback for verification and validation in particular data pre-processing techniques. The feedback is required from the main data analysis process. The complete process is associated with good job if the processed data provides valuable information in the later of data analysis process.

Ladha et al. in [11] have mentioned following advantages of feature selection:

1. Reduce the dimensionality to reduce storage requirements and increase algorithm speed.
2. Remove redundant, irrelevant or noisy data.
3. Improve data quality.
4. Increase the accuracy of resulting model.
5. Feature set reduction.
6. Improve the performance to gain predictive accuracy.
7. Gain knowledge about the process.

The objective of feature selection and feature extraction concerns the dimension reduction to improve analysis of data. The important aspect becomes relevant on considering real world datasets having hundreds or thousands of features. The author concluded to first perform the reduction without changing them whereas feature extraction reduces dimensionality. It emphasised to create other features that was more significant by computing transformation of the unique features [12].

Convolutional neural network (CNN) is a type of deep learning neural network (NN) for image recognition and classification. The link has valued weights that are finely tuned in the training process that results in a trained network. The neural network has layers connected by artificial neurons. Each layer categorizes a set of simple patterns of the input. A standard CNN have 5–25 layers that ends with an output layer [13]. A large database of good quality images with strong and distinctive features is taken for good classification results. The CNN model on classification accuracy removes the need of a Graphics Processing Unit (GPU) for training despite the advantage of shortening the training time and the transfer learning exceeds full training [14].

Automated behaviour-based malware detection using machine learning techniques is a thoughtful and fruitful method. It has generated behaviour reports on an emulated (sandbox) environment and the malware will be identified automatically. Different classifiers namely k-Nearest Neighbours (kNN), Naïve Bayes, J48 Decision Tree, Support Vector Machine (SVM), and Multilayer Perceptron Neural Network (MLP) were used. The experimental results achieved by J48 decision tree with a recall of 95.9%, a false positive rate of 2.4%, a precision of 97.3%, and an accuracy of 96.8% concluded that the use of machine learning techniques detects malware quite efficiently and effectively [15].

Researchers [16] have proposed a data-driven methodology consisting of CI methods to compare and forecast AQ parameters. The PCA and ANNs methods were chosen for the CI tasks. Implementation of a novel hybrid method for selecting the input variables of the ANN-MLP models were considered. To improve the accuracy of the forecasting model a nonlinear algorithm such as the ANN-MLP, the multi-fold training and validation scheme were adopted for air pollution.

The four algorithms of supervised learning, were compared for calculating correctness rate of waste generation. For successfully forecasting the future trends in MSW generation deep learning and machine learning models provide promising tools that may allow decision makers for planning MSW management purposes. They concluded that ANFIS, kNN, and SVM models results best prediction performance. In addition, results suggest that ANFIS model produced more accurate forecasts than kNN and SVM. Hence kNN modelling is applied for waste generation forecasting [17].

Reference [18] revealed the importance of laying a sustainable foundation for advanced technologies in intelligent buildings based on green Architecture and Integrated Project delivery. They concluded that it should be done before Nanotechnology, Building Information Modelling and Lean Construction along with Artificial Intelligence to achieve intelligent buildings and concern global warming.

3 Methodology

In this paper, supervised learning techniques have been used to propose a method which can help us to understand the air quality index of different states of India and create a model to predict the quality of air on the basis of presence of Nitrogen Oxide (NO_2) and Particulate Matter (PM10) on the basis of understanding and knowledge. It can also help us to reduce the area of interest which requires urgent attention in terms of air quality. Furthermore this model can be implemented to reach a larger goal of worldwide air quality prediction. We have used supervised learning to make a prediction model for air quality index of an area by gathering a basic statistical information. Through similar supervised learning methodology in deep learning image classification, another method has been proposed for effective waste management by expanding the scope of deep learning into other fields like Internet of things [19–21]. The error approximations used in this paper to predict the accuracy of the machine learning models are, Mean Absolute Error and R Squared error.

Supervised Learning for Predictive Analysis

In the following case a government dataset has been collected from government source site regarding Nitrogen Dioxide (NO_2) content and particulate matter (PM10) content present in the air in different states of India for the year 2011. The presence of these contents play a very important role in determining the air quality [22, 23].

Steps Involved:

1 Gathering Raw Data
2 Extracting Desired Data
3 Pre-process Data
4 Visualise Data
5 Comparing Data and Knowledge Points
6 Extracting Right Information
7 Creating a model for predictive analysis using machine learning
8 Cross validating a model with future data.

Step 1:

URL: data.gov.in

Data was collected through website scraping and open data organisations like data.gov.in. The mentioned URL source is an authentic government website which provides legitimate data collected by the government of India on different topics such as air quality, water quality etc. [24].

In this paper, we have used data associated to the statistics of nitrogen dioxide and particulate matter content present in the air for the year 2011and 2012. For data visualisation purposes in order to find knowledge points the data of both NO_2 and PM10 are considered. Also the data of PM10 is verified and consistent to be analysed. The analysis suggests that the control of PM10 in comparison to PM2.5 levels in subway systems of India should be promoted to significantly emphasise on economic benefits. Here Machine Learning process was based on the content of PM10 for the year 2011 for training and content of PM10 for year 2012 for testing [25].

Step 2:

Data extraction is the process of reading the valuable data from raw data file including that of csv, sql etc. The result analysis is being done using program generated graphs in Bohek and Matplotlib libraries in python. Figure 1 represents the initial data before performing data cleaning and data engineering process. In Fig. 1, we have taken suitable columns with their raw data which was most desirable to perform feature selection process as per our needs.

Step 3:

Data needs to be pre-processed to fit for visualisation purposes.

```
data_nitrogen = pd.read_csv('air_quality_nitrogen_dioxide_2011.csv',encoding='latin1')
nitrogen_data = data_nitrogen.copy()
nitrogen_data = nitrogen_data.drop(['City','Location', 'Station code', 'Type', 'Category of ES'], axis=1)
nitrogen_data['Air Quality'].unique()
nitrogen_data['Air Quality Index'] = nitrogen_data['Air Quality'].map({'Low':1, 'Moderate':2, 'High':3, 'Critical':4})
nitrogen_data.head()
```

	State	No. of mon. days (n)	Min	Max	NO2 Annual average (µg/m3)	10 percentile	90 percentile	Std. Dev.	Percentage- exceedence(24 hourly)	Air Quality	Air Quality Index
0	Andhra Pradesh	72	9	9	9	9	9	0	0	Low	1.0
1	Andhra Pradesh	69	9	13	10	9	11	1	0	Low	1.0
2	Andhra Pradesh	94	5	55	33	24	40	8	0	Moderate	2.0
3	Andhra Pradesh	93	13	54	32	21	38	7	0	Moderate	2.0
4	Andhra Pradesh	95	10	62	35	27	41	7	0	Moderate	2.0

Fig. 1 Representation of raw data from file

Step 4:

Data needs to be understood before taking out the valid knowledge points required to perform knowledge engineering [26]. Visualisation of data makes the process of data engineering and knowledge extraction a convenient process. The data shown below graphs shows the air quality against each state of India on a scale of 1-4 where 1 is the category of good conditions while 4 stands for a highly critical situation (Figs. 2 and 3).

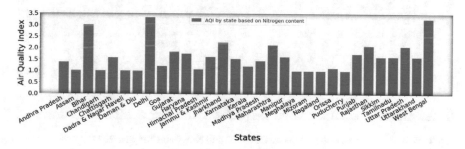

Fig. 2 AQI of Nitrogen Dioxide (NO$_2$) content on critical scale of 1–4

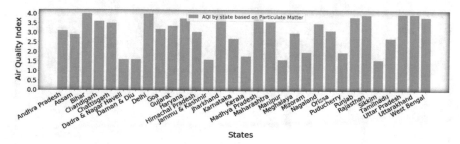

Fig. 3 AQI of Particulate Matter (PM10) on critical scale of 1–4

Step 5:

The bar graph represented in Fig. 4, clearly represents level of air quality in different states. This visualisation is very helpful in choosing the area of interests which should be on high alert to work on air quality control.

It is evidently observable that highly polluted states include Delhi, West Bengal and Bihar which have the most critical air conditions on the basis nitrogen dioxide and particulate matter. This information can be used by major air quality control agencies to reduce their work effort towards the areas that are highly polluted.

In Fig. 5, it is shown that by data visualisation it is possible to identify the states of India which have a very high content of nitrogen dioxide and particulate matter.

In the succeeding steps, a predictive model has developed to predict the quality of air on the basis of information regarding a particular hazardous content. This process of data visualisation can be used to narrw down the area.

Step 6:

A machine learning model should be trained with right data to avoid any case of under fitting or over fitting of data [26–28]. Since the data which was obtained from

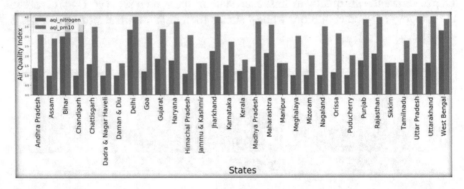

Fig. 4 State wise comparison for content of PM10 and NO_2 in air for year 2011

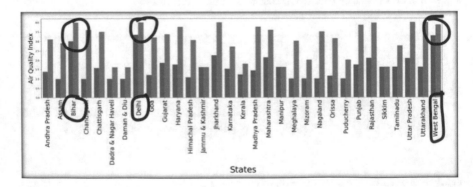

Fig. 5 States with low air quality index of PM10 and NO_2

	State	City	Location	Station Code	Type	Category of ES	No. of mon. days (n)	Min	Max	PM10 Annual average (µg/m3)	10 percentile	90 percentile	Standard Deviation	Percentage-exceedence(24 hourly)	Air Quality
0	Andhra Pradesh	Chitoor	GNC Toll Gate Tirumala	582	RIRuO	NaN	103	24	52	40	35	45	4	0	Moderate
1	Andhra Pradesh	Guntur	Near Hindu College, Market Road	583	RIRuO	NaN	105	71	79	75	2	73	77	0	High
2	Andhra Pradesh	Hydrabad	Tarnaka, NEERI Lab. IICT Campus	150	RIRuO	NaN	88	8	70	26	17	35	10	0	Low
3	Andhra Pradesh	Hydrabad	Nacharam, Industrial Estate	151	RIRuO	NaN	87	12	100	30	19	42	15	0	Low

Fig. 6 Feature set before performing feature correlation and feature selection

the government resources was very clean in itself. Thus to achieve the accurate fitting of model concept of dimensionality reduction was applied by simply removing the irrelevant data.

Figure 6 shows the representation without performing any correlation. It is inferred that State, City, Location and Station Code will only result into over fitting of data because it hold no scientific relationship to chemical composition of components of air.

In Fig. 7 represents a final feature list has been chosen on the basis of statistical feature selection and correlation. Furthermore, this feature was originally named as "Air Quality" and that has been renamed to "Air Quality Index". The data type of this feature is changed from integer to string for better classification and more relatable content in terms of intra data types.

Step 7:

After feature selection a thorough study of data for predictive model is designed which could classify the air quality of an area in terms of presence of particulate matter content in the particular area. A mixed regression-classification approach was used. There are four labels of classification on the basis of quality including:

1. Good—1
2. Moderate—2
3. High—3
4. Critical—4.

where 1 indicates the best quality of air.

Step 8:

Model Approach Used:

1. All the location based columns and strictly string columns were dropped.

No. of mon. days (n)	PM10 Annual average (µg/m3)	10 percentile	90 percentile	Standard Deviation	Percentage- exceedence(24 hourly)	Air Quality Index

Fig. 7 Final feature set after performing feature correlation and feature selection

2. Filling the empty column values of air quality index with the average of all the available column values of that particular state.
3. In case all the column values of a state are empty, those column values were filled with average of all the column values of air quality index.
4. All the other column values were filled by 0 default because the collected data already contained all the necessary statistical data necessary for prediction.
5. Data for air quality on the basis of particulate matter was splits into testing and training sets. The data used for this purpose was for the year 2011.
6. The model was trained on a regress or algorithm using Random Forest Regression algorithm on training data for 2011.
7. A k-fold cross validation was used to verify model on different splits.
8. Finally the model was tested on test data for 2011. The model was able to reach an accuracy of 84% approx. as shown in Fig. 8.
9. Then the regression predictions were made on 2012 data which was introduced to the model for the very first time. In this approach, a R^2 square of 0.88 was achieved and Mean Absolute Score of 0.10 was achieved as represented in Fig. 9.
10. To fine tune this result further, a classification approach was taken where the predicted output is rounded off to reduce the distance between actual predictions and proposed predictions. After this algorithm tuning both R^2 scores and Mean Absolute Error were improved as shown in Fig. 10.

```
predict = regr.predict(X_test)
regr.score(X_test, y_test)

0.8440907077825055
```

Fig. 8 Accuracy of predictive model on the basis of testing data

```
from sklearn.metrics import r2_score
print("R2 score : %.2f" % r2_score(final_predict,
    compare, multioutput='variance_weighted'))

R2 score : 0.88

from sklearn.metrics import mean_absolute_error
print("Mean Absolute Score : %.2f" %
    mean_absolute_error(final_predict, compare))
print('Mean Absolute Error:',
    round(np.mean(errors), 2), 'degrees.')

Mean Absolute Score : 0.10
Mean Absolute Error: 0.05 degrees.
```

Fig. 9 Error analysis on the basis of regression approach

```
from sklearn.metrics import r2_score
print("R2 score : %.2f" % r2_score(final_predict,
    new_predict, multioutput='variance_weighted'))
```

R2 score : 0.92

```
from sklearn.metrics import mean_absolute_error
print("Mean Absolute Score : %.2f" %
    mean_absolute_error(final_predict, new_predict))
print('Mean Absolute Error:',
    round(np.mean(errors), 2), 'degrees.')
```

Mean Absolute Score : 0.05
Mean Absolute Error: 0.05 degrees.

Fig. 10 Error report on the basis of regression to classification approach

For evaluating the accuracy using mean absolute error, R^2 and Mean Absolute Score has been determined. The mentioned figures are further extended to create a reportable accuracy percentage using mean absolute error accuracy. This accuracy has been derived by getting the absolute difference between predicted values and actual values. This absolute value list was further used to find the accuracy percentage of proposed model (Fig. 11).

It has been observed that the proposed approach was able to achieve 96% accuracy approximately on a new dataset which was never introduced to the model while training it.

Advantages towards reaching the goals of sustainable development:

1. Data Engineering and Knowledge Engineering applied through above example is a very useful tool to spot the area of concerns which needs urgent attention to reach the goals of sustainable development.
2. The predictive analysis can work independent of sensing electronic devices to predict the future of trends in climate change by learning the previous trends.
3. Highly cost effective method and it requires very less human intervention once it has being trained with enough data.
4. Can be easily extended to multiple devices synchronously to gather information.

This was just one of the many use cases where machine learning, a subset of artificial intelligence, can be used to understand the area of concern which are becoming a deterrent towards reaching the goals of sustainable development.

Fig. 11 Accuracy predictions on untrained dataset

```
errors = abs(new_predict - final_predict)
mape = 100 * (errors / final_predict)
accuracy = 100 - np.mean(mape)
print('Accuracy:', round(accuracy, 2), '%.')
```

Accuracy: 96.75 %.

IoT and Deep Learning for Surveillance

Deep learning provides a very powerful algorithms for image identification and classification which can be combined with the certain other technologies to carry out and reach the goals of sustainable development on small, medium and large scales very efficiently [27].

Convolutional Neural Network (CNN)

A CNN is a deep learning model based methodology that is primarily used to classify images and objects through similar cluster. Deep learning convolutional neural networks algorithms works on the concepts of learn and classify on the basis of experience. Apart from image classification convolutional neural network plays an important role in text classification and text analytics.

A CNN usually consists of these basic layers to draw a conclusion for text analysis:

- Input Layer
- Hidden Layer
- Pooling Layer
- Max Pooling Layer
- Conv + RELU
- Output Layer (Classification Layer)
- Flatten Layer
- Fully Connected Layer
- Soft Max Layer.

Internet of Things (IoT)

Internet of things is a field of engineering and computer science in which multiple things are interconnected through WLAN of internet. In the simplest of sense IoT is a set of electronic devices which are continuously communicating with each other to carry out a certain task to reach a final output.

Proposed Method to Achieve Sustainable Development Goals:

1. Clean Water and Sanitation
2. Good Health and Well Being
3. Life Below Water
4. Life on Land
5. Architectural Components
6. Deep Learning for Image Classification
7. Drones for remote surveillance and observation
8. Analyser for training more data.

Steps Involved:

1. Train an offline CNN model for image classification of different types of material waste.

2. Writing a program to control a drone using a laptop and receive real time image captured through drone camera.
3. Interfacing drone to store image data.
4. Feeding the image data to the CNN classifier to get regular updates of the types of trash around an area.

4   Experimental Results

In the following result set we demonstrate how machine learning and deep learning can be used to reach the goals of sustainable development by using the mentioned techniques for effective waste management. Every day waste is generated in metric tons [29] which, if not managed properly is very hazardous to planet and endangers human population at large. Figure 12 shows how neural network can be trained to classify different types of waste materials not scientifically but generally which could be easily understood by a layman. Figures 13, 14, 15 and 16 shows the practical implementation of same neural network where 13 and 14 are test images on which the classifier was trained and Figs. 15 and 16 are images and test results of completely new image meaning they were not introduced to neural network while training of data.

Once this neural network is in place, then it can be used to further extend to an IoT device to classify dump on the fly or in real time. Figure 17 introduce a basic idea or architecture while Fig. 18 expands the capability of this architecture which is more efficient and more economical in long run. (Those architecture are ideas which we have proposed)

Step 1: Setting up an offline image classifier.
CNN Classifier Used: InceptionNet v3

Convolutional neural networks interprets an image as a set of numbers. Each image fed into this network goes through several layers which are used to extract distin-

```
INFO:tensorflow:2300 bottleneck files created.
INFO:tensorflow:2400 bottleneck files created.
INFO:tensorflow:2500 bottleneck files created.
WARNING:tensorflow:From retrain.py:790: softmax_cross_entropy_with_logits (from
Instructions for updating:

Future major versions of TensorFlow will allow gradients to flow
into the labels input on backprop by default.

See `tf.nn.softmax_cross_entropy_with_logits_v2`.

INFO:tensorflow:2019-03-06 17:01:07.998185: Step 0: Train accuracy = 36.0%
INFO:tensorflow:2019-03-06 17:01:07.998491: Step 0: Cross entropy = 1.866903
```

Fig. 12 Training of the neural network model for image classification

Fig. 13 Test Case 1 Image
fed into neural network

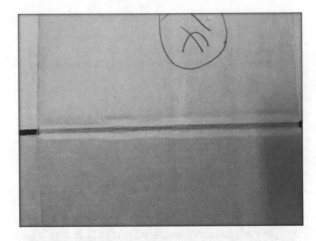

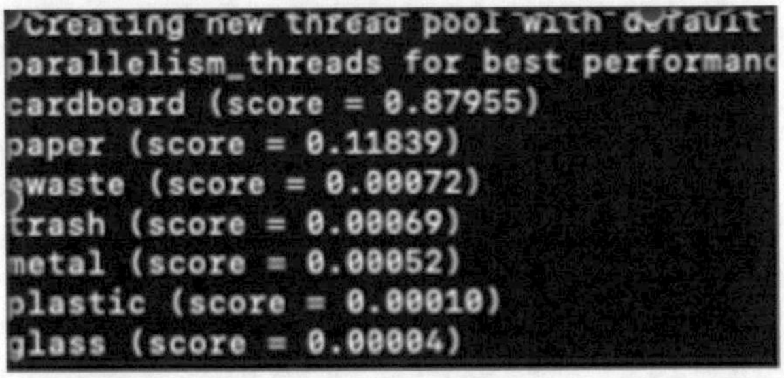

Fig. 14 Results of Test Case
1 prediction on trained model

```
Creating new thread pool with default
parallelism_threads for best performanc
cardboard (score = 0.87955)
paper (score = 0.11839)
ewaste (score = 0.00072)
trash (score = 0.00069)
metal (score = 0.00052)
plastic (score = 0.00010)
glass (score = 0.00004)
```

Fig. 15 Test Case 2 Image
fed into neural network on
which the model was not
trained, for classification

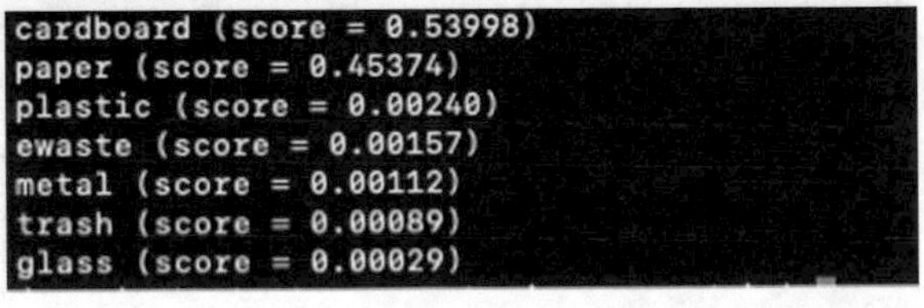

Fig. 16 Results of Test Case
2 prediction

```
cardboard (score = 0.53998)
paper (score = 0.45374)
plastic (score = 0.00240)
ewaste (score = 0.00157)
metal (score = 0.00112)
trash (score = 0.00089)
glass (score = 0.00029)
```

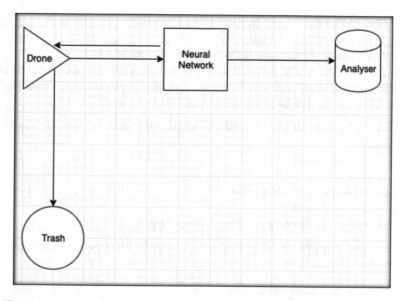

Fig. 17 Basic architecture of the proposed model for waste surveillance

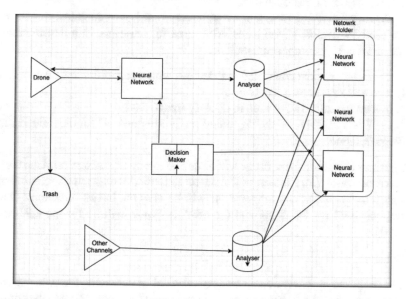

Fig. 18 Advanced architecture of the proposed model for waste surveillance

guished features at each layer. In case of Google's Inception Net at the first layer pooling and max pooling of a image is performed. Training is done using inception model which automatically extracts the desired features of an image.

After training the neural network with image data, a manual test is run to check the validation and authenticity of the model for precise predictions. There are two test cases generated to cross check the validation of this model.

Test Case 1: Testing for image database which was used to train the model for classification.
Actual Image: cardboard.
Predicted Result: Cardboard 87.95% (Fig. 14).
Test Case 2: Image Taken On the Spot.
Actual Image: cardboard.
Predicted Image: cardboard with 54% accuracy approx.

This shows that our model is accurate enough to be reused and extended for classification of trash.

Step 2: Configuration the IoT Device

Step 3: Frame Capture and Analyser
The following Fig. 17, depicts a basic architecture of how a deep learning network can be used for remote surveillance of an area for trash tracking and monitoring around an area. Key components include:

1. Drone or any remote controlled IoT devices with a capability to take continuous frames and send it over a network.
2. A trained neural network used to classify image data.
3. An analyser to store and the captured images are fine tune for the trained neural network model.

The camera attached to the drone will be continuously capturing the image frames. The type of waste from image will then be classified using the captured frames through neural network and the response will be send out in two directions. One of the response will for the drone and the other response will be for the analyser to collect data.

Advantages:

1. The proposed architecture can be very efficient in surveying the types of waste around an area which are not accessible to human otherwise. Due to the compact size of drones and ability to control remotely reaching remote areas becomes a much easy task.
2. By making a small submarine equivalent of a drone, underwater waste management can reduce the amount of money spent on underwater waste surveillance.
3. The proposed method can exponentially reduce the effort time and cost of human labour and can quickly be used to determine area of interests for effective waste management.

Extending the scope of neural network for local waste education:

The neural network model trained in the above proposed method can be expanded to create a larger centralised network for garbage and trash monitoring and removal for cleaner areas [30].

Components of Extension:

The proposed architecture can be extended to work more efficiently and precisely by fine tuning the basic architecture. The components include:

1. Drone with a capability to pick up objectionable garbage or trash object through the addition of bionic arm or robotic parts. Since drone comes in various sizes and weights, they are capable of performing robotic actions.
2. Use of multiple channels to capture images. Multiple channels include smartphones and web scrapers which can be used as a multipurpose image providers and educators on waste management.
3. An active neural network will play continuously providing result to all channels and drone regarding the classification of an image.
4. An offline neural network subsystem will be a set of multiple neural networks which are being continuously provided with images received from multiple analysers and each of the continued neural network is vigorously trained on received images.
5. Multiple neural networks will be continuously provided with images received by analyser. A human interface can sit at analyser level for couple of hours a day in initial days to distinguish between false positive and false negatives from active neural network layer.
6. Multiple Analysers will be set of systems holding offline image data. They will be pushing image data towards the neural network subsystem container for training of data. Analysers will be performing the task of digital image processing if and when necessary and other decisions regarding pushing and pulling image data between active layer and offline layers of neural network.
7. A decision maker will be sitting between active and inactive layers of neural network. As soon as one of the neural network in inactive layer gains an accuracy higher than currently active layer, the decision maker will switch the more experienced inactive layer as active layer will pushing the lower experienced neural networks towards retraining.

Advantages of the proposed architecture model for waste classification and management:

1. Automatic trash removal.
2. Higher accuracy and hit rate for more accurate predictions.
3. Use of multiple channels to gather data.
4. More effective way to reach masses regarding efficient waste classification and management.

5 Conclusion and Future Work

Sustainable development has become a topic of utmost importance and a concern due to multiple environmental hazards. It has become necessary that we realise the importance of resource utilization and management without making a dent in the economy. The methods and techniques explained here is pointing towards achieving the goals of sustainable development by leveraging the power of artificial intelligence technologies like machine learning and deep learning.

1. Right Data can be used to learn the area of interest which needs attention to achieve the goals of sustainable development.
2. Data Engineering and gathering precise data points is highly inevitable in developing machine learning models which is further used to make an economically feasible product such as air pollution predictor.
3. Predictive data analysis can be used to timely predict hazardous weather patterns and for future mitigation strategy.
4. Deep Learning can be used for effective waste management.

With extensive work being put in the field of Artificial Intelligence machine learning, deep learning, computer vision etc. sustainable development looks like an achievable milestone. By 2030, which is the target year of United Nations for achieving all goals of sustainable development. AI would have found a lot of new methods and technology and it will play an important role in solving the more complex use cases of Sustainable development.

References

1. Chichilnisky G (2000) An axiomatic approach to choice under uncertainty with catastrophic risks. Resour Energy Econ 22(3):221–231
2. Lele SM (1991) Sustainable development: a critical review. World Dev 19(6):607–621
3. Robert KW, Thomas MP, Leiserowitz AA (2005) What is sustainable development? Goals, indicators, values, and practice. Environ: Sci Policy Sustain Dev 47(3):8–21
4. Yun, JH et al (2016) Not deep learning but autonomous learning of open innovation for sustainable artificial intelligence. Sustainability 8(8):797
5. Dean J, Corrado G, Monga R, Chen K, Devin M, Mao M, Ng AY (2012) Large scale distributed deep networks. In: Advances in neural information processing systems, pp 1223–1231
6. Spring R, Shrivastava A (2017) Scalable and sustainable deep learning via randomized hashing. In: Proceedings of the 23rd ACM SIGKDD international conference on knowledge discovery and data mining. ACM
7. Michie D, Spiegelhalter DJ, Taylor CC (1994) Machine learning. Neural Stat Classif 13
8. Studer R, Benjamins VR, Fensel D (1998) Knowledge engineering: principles and methods. Data Knowl Eng 25(1–2):161–197
9. Kendal SL, Creen M (2007) An introduction to knowledge engineering. Springer, London
10. Famili A et al (1997) Data preprocessing and intelligent data analysis. Intell Data Anal 1(1):3–23
11. Ladha L, Deepa T (2011) Feature selection methods and algorithms. Int J Comput Sci Eng 3(5):1787–1797

12. Khalid S, Khalil T, Nasreen S (2014) A survey of feature selection and feature extraction techniques in machine learning. In: 2014 science and information conference. IEEE
13. Hijazi S, Kumar R, Rowen C (2015) Using convolutional neural networks for image recognition. Cadence Design Systems Inc., San Jose
14. Chin CS et al (2017) Intelligent image recognition system for marine fouling using Softmax transfer learning and deep convolutional neural networks. Complexity 2017
15. Firdausi I, Erwin A, Nugroho AS (2010) Analysis of machine learning techniques used in behavior-based malware detection. In: 2010 second international conference on advances in computing, control, and telecommunication technologies. IEEE
16. Voukantsis D et al (2011) Intercomparison of air quality data using principal component analysis, and forecasting of PM10 and PM2. 5 concentrations using artificial neural networks, in Thessaloniki and Helsinki. Sci Total Environ 409(7):1266–1276
17. Abbasi M, El Hanandeh A (2016) Forecasting municipal solid waste generation using artificial intelligence modelling approaches. Waste Manag 56:13–22
18. Adio-Moses D, Asaolu OS (2016) Artificial intelligence for sustainable development of intelligent buildings. In: Proceedings of the 9th CIDB postgraduate conference. University of Cape Town, South Africa
19. Lary DJ, Lary T, Sattler B (2015) Using machine learning to estimate global PM2. 5 for environmental health studies. Environ Health Insights 9(EHI-S15664)
20. Li X, Peng L, Hu Y, Shao J, Chi T (2016) Deep learning architecture for air quality predictions. Environ Sci Pollut Res 23(22):22408–22417
21. Chelani AB (2010) Prediction of daily maximum ground ozone concentration using support vector machine. Environ Monit Assess 162(1–4):169–176
22. https://www.who.int/news-room/fact-sheets/detail/ambient-(outdoor)-air-quality-and-health. Accessed on Sept 2018
23. Vallero DA (2014) Fundamentals of air pollution. Academic press
24. Al-Jarrah OY, Yoo PD, Muhaidat S, Karagiannidis GK, Taha K (2015) Efficient machine learning for big data: a review, big data research, vol 2, no 3, pp 87–93, ISSN 2214-5796. https://doi.org/10.1016/j.bdr.2015.04.001
25. Madu CN, Kuei C, Lee P (2017) Urban sustainability management: a deep learning perspective. Sustain Cities Soc 30, 1–17, ISSN 2210-6707. https://doi.org/10.1016/j.scs.2016.12.012
26. Warburton K (2003) Deep learning and education for sustainability. Int J Sustain High Educ 4(1):44–56
27. Gassmann O, Enkel E (2004) Towards a theory of open innovation: three core process archetypes
28. Sun R, Peterson T (1998) Autonomous learning of sequential tasks: experiments and analyses. IEEE Trans Neural Networks 9(6):1217–1234
29. Sutton RS, Barto AG (1998) Reinforcement learning, a Bradford book. MIT Press Cambridge, MA 2015(3):2
30. Lam KC, Lee D, Hu T (2001) Understanding the effect of the learning–forgetting phenomenon to duration of projects construction. Int J Project Manage 19(7):411–420

Brain Tumor Segmentation Using OTSU Embedded Adaptive Particle Swarm Optimization Method and Convolutional Neural Network

Surbhi Vijh, Shilpi Sharma and Prashant Gaurav

Abstract Medical imaging and deep learning have tremendously shown improvement in research field of brain tumor segmentation. Data visualization and exploration plays important role in developing robust computer aided diagnosis system. The analysis is performed in proposed work to provide automation in brain tumor segmentation. The adaptive particle swarm optimization along with OTSU is contributed to determine the optimal threshold value. Anisotropic diffusion filtering is applied on brain MRI to remove the noise and improving the image quality. The extracted features provided as data for training the convolutional neural network and performing classification. The proposed research achieves higher accuracy of 98% which is better over existing methods.

Keywords Brain tumor · Segmentation · Adaptive particle swarm optimization · Deep learning · Deep neural network · Convolutional neural network

1 Introduction

Artificial intelligence and computer vision are showing remarkable improvement in development of automatic medical diagnosis system. According to National Brain Tumor Society, it is estimated that 78,980 new cases are predicted in 2018 in the United States. Cancer has been observed and considered as the disease that may

S. Vijh
Department of Information Technology, KIET group of Institutions, Ghaziabad, India
e-mail: surbhivijh428@gmail.com

S. Sharma (✉) · P. Gaurav
Department of Computer Science and Engineering, Amity School of Engineering and Technology, Amity University Uttar Pradesh, Sector 125, Noida, India
e-mail: ssharma22@amity.edu

P. Gaurav
e-mail: prashantgaurav36@gmail.com

© Springer Nature Switzerland AG 2020
J. Hemanth et al. (eds.), *Data Visualization and Knowledge Engineering*,
Lecture Notes on Data Engineering and Communications Technologies 32,
https://doi.org/10.1007/978-3-030-25797-2_8

cause maximum death around the world as it is difficult to perform diagnosis of patient. The Brain cancerous tumor can broadly be classified [1] into primary tumor and secondary tumor (metastatic tumor). Image processing techniques [2] are applied in detection and exploring the abnormalities issues in tumor detection. The early stage of detection can improve the process of decision making by radiologist, physician, experts and can increase the survival rate of Tumor patients. The automatic computer aided diagnosis system is proposed in medical imaging with the usage of deep learning. Brain tumor occurs due to the presence of abnormality as the cell divides and its growth increase. The different modalities of Magnetic resonance imaging are present such as T1-weighted MRI, T2-weighted MRI, Flair and Flair with contrast enhancement. The segmentation of the brain tumor MRI is the most challenging task in contributing towards the development of diagnosed system [3]. A new approach is proposed in construction of automatic diagnosis system. The simulated steps performed in proposed system are (a) image acquisition (b) pre-processing (c) segmentation (d) post processing (e) feature extraction (f) classification. There are various segmentation [4] approaches used previously such as region growing technique, Thresholding, clustering, edge detection etc. In the proposed work, the novel hybrid approach of OTSU and adaptive particle swarm optimization is adopted along with convolutional neural network. Nature inspired metaheuristic optimization algorithm are used for optimizing the problem by resembling the physical or biological phenomenon. In computer science, there are various metaheuristic optimization technique such whale optimization algorithm [5], Particle swarm optimization [6], Adaptive PSO, Ant bee colony [7], Ant colony optimization (ACO), cuckoo search [8]. The adaptive particle swarm optimization technique is implemented because of improved time and space performance. The OSTU + APSO approach is providing the optimal threshold value which enables to perform the segmentation in improved manner. Anisotropic diffusion filtering is applied for smoothing and denoising the brain MRI. Feature extraction is performed with the usage of GLCM technique in which all statistical and texture characteristics are extracted. The data organized in feature extraction stage is used for training and testing of convolutional neural network for classification. The classification is performed to predict whether the tumor is present or not. Deep neural network considers the selection of network structure, activation function, number of neurons, learning rate and other factors which are highly dependent on data representation [9]. Convolutional neural network contributing towards the medical imaging data visualization, analyzation and exploration of large amount of data. Convolutional neural network enables to design the computer aided diagnosis system with higher accuracy in medical image domain [10]. The performance of the developed automatic computer diagnosis system is measured through parameter accuracy. The accuracy obtained of proposed system is 98% which is better than any other existing system.

The proposed paper is organized into different section in which Sect. 2 describes regarding literature review and its related work, Sect. 3 represents the proposed methodology and contribution towards the paper, Sect. 4 presents the simulated experimental results, Sect. 5 describes related to result and its discussion, Sect. 6 states the conclusion and future scope of the proposed work.

2 Literature Review

In recent years, the research in medical imaging is increasing exponentially to show advancement in data visualization and improving the survival rate of patients at early stage. The objective of researchers is to develop the robust computer aided automatic diagnosis system for detection of brain tumor. The emerged growing technique known as deep learning is contributing to pattern recognition, artificial intelligence, machine learning applications [11–13]. The computational architecture containing multiple processing layers are trained to acknowledge the data visualization and its various representation [14]. Convolution neural network is considered promising method in fields of computer vision, various application of medical imaging analysis and natural language processing. The deep learning techniques helps in providing effective improved performance and approaches [15, 16]. Medical diagnosis is the most challenging decision-making process. The fundamental procedure of artificial intelligence helps the radiologist and experts in taking decision regarding the process of treatment [17, 18]. The deep learning outperforms in handling the immense data images and the multiple layer of abstraction. The convolutional neural network can be used to determine wider range of perceptual task and object recognition [19].

There are numerous technique which are used to obtain information of human organ among which MRI is desired non-invasive method used to evaluate the neural activity of human brain [20]. MRI was invented in 1970 considered as medical imaging technique providing the internal details as well essential information of irregularity and contrast changes in human soft tissue [21]. The magnetic properties of tissues are observed using magnetic field, radio waves of MRI scanners and produces the images having structural analysis of human body. Magnetic resonance imaging lead emphasis on visualization of tissues accommodating hydrogen such as Cerebrospinal fluid (CSF), brain, bone marrow etc. The MRI images are interpreted and analyzed by radiologist, experts, physician to detect and screen the abnormality present in brain [22]. Neoplasm (tumor) are broadly structured into three types (a) Benign (b) Pre-malignant (c) Malignant [23].

Medical image processing has become very efficient technique for treatment of cancer patient and detection of tumor. Brain Tumor segmentation [24] are broadly classified into three categories (a) manual (b) semiautomatic (d) fully automatic. The manual segmentation is performed by radiologist and experts which provides time consuming and poor outcomes. Thus, semiautomatic is introduced in which human know few parameters on basis of which analyzation is performed. It provided better results than manual segmentation. However, fully automatic segmentation system is emerged as booming technique which outperforms and works on the basis of prior information using the concept of artificial intelligence. The different modalities of tumored brain MRI shown in Fig. 1.

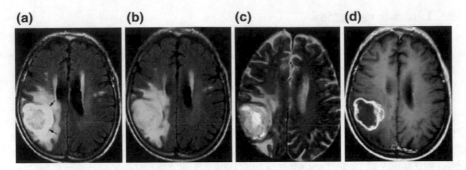

Fig. 1 **a** T1-weighted MRI **b** T2-weighted MRI **c** Flair **d** Flair with contrast enhancement

Image segmentation of brain magnetic resonance imaging considered as difficult and important process for detection of brain tumor and extracting the important features for determining the abnormalities present [21]. It is very difficult task needed to be performed since the abnormal image holds various irregularities in shape, structure and size and location [25]. The selection of right segmentation technique helps to perform skull stripping of brain MRI. The skull stripping is used to remove the cerebellum tissues on different modalities of images such as T1 weighted and T2 weighted magnetic resonance image in various medical applications [26, 27]. Patel et al. [28] presented different segmentation techniques for studying the detection of brain tumor named as (a) edge-based method (b) Region growing [29] (c) clustering method (d) Fuzzy c-mean clustering [30] (e) Thresholding. Cabria et al. [31] proposed a new technique for segmentation named as fusion of potential field clustering and another ensemble method. The methodology is applied on BRATS MRI benchmark for detection of tumor region. Ayachi et al. [32] developed automatic computer aided diagnosis system for brain tumor segmentation. The pixel classification segmentation technique is applied and the system involves support vector machine for classification Gliomas dataset. Soleimani et al. [33] proposed methodology in which ant colony optimization technique is considered for determining brain tumor segmentation and improving the accuracy. The usage of metaheuristic techniques in development of diagnosis system helps to obtain optimized threshold value of parameters. Jothi et at [34] stated a novel approach for optimal feature selection of brain MRI using firefly based quick reduct and tolerance rough set. Manic et al. [35] implemented the multilevel thresholding based upon firefly algorithm to segment the gray scale image by using kapur/Tsallis entropy to find optimal threshold of image and then performance is computed on parameters such as root mean squared error, Normalized absolute error, Structural Similarity Index Matrix, Peak signal to noise ratio (PSSR). Sharma et al. [36] proposed a methodology in which statistical features of MRI brain image are extracted using GLCM. The k-mean and ANN model is created using the extracted information for determining the performance parameters.

Jafari et al. [37] presented a novel approach for developing automatic tumor detection system of brain and classifying whether brain image is normal or abnormal. The genetic algorithm is implemented for considering the selected optimal features. The classification technique used for detection is support vector machine achieving accuracy up to 83.22%. Jiang et al. [38] presented the medical image analysis and characterization with the usage of intelligent computing. The medical image segmentation and edge detection are formulated with the support of different artificial neural network models such as feedforward neural network, feedback neural network, self-organizing maps to obtain the medical image analysis. They are shown the various application where ANN can be used extensively such as in tumor tracking [39], image compression [40] and enhancement [41].

Havaei et al. [42] proposed a methodology for automatic brain tumor segmentation of glioblastomas MRI using deep learning. The various convolutional neural network architecture is designed among which fully connected CNN is used for determining the analysis on BRATS (2013) testing dataset. It is evaluated and observed that the proposed methodology was providing improved accuracy, speed over traditional methods in computer vision. Gao et al. [43] stated a novel approach by considering the brain CT images to perform classification by using deep learning techniques. The diagnosis computer aided system is developed for predicting the Alzheimer's disease. They partitioned the dataset into three parts AD, lesion and normal ageing and predicted the classification using 2D and 3D CNN networks architecture. The average of 87.6% of accuracy is obtained for the categorized structure of images.

Pereira et al. [44] performed automatic segmentation of brain tumor in MRI images by considering BRATS 2013 And BRATS 2015 dataset. The technique used is convolutional neural network for data visualization and analysis. The system is validated using BRATS 2013 database and acquired first position in dice similarity coefficient matrix for the target dataset. Sharma et al. [45] designed a methodology using differential evolution embedded OTSU and artificial neural network for the automatic segmentation of brain tumor. The brain MRI of 58 patients are considered for evaluating the performance parameters. The accuracy of proposed system obtained is of 94.73%.

Mohan et al. [46] showed the analysis on medical image of brain tumor magnetic resonance imaging and its grade classification is performed. The hybrid approach of image processing, machine learning and artificial intelligence is used to improve the accuracy of diagnosis system. The methodology involves extraction and performing grading of tumor. Chen et al. [47] proposed a novel approach of segmentation on 3D MR image of brain and used Voxel wise residual network (VoxResNet) of deep learning to obtain volumetric performance information of image. Zhao et al. [48] presented brain tumor segmentation model with the usage of hybrid network of fully connected convolutional network and conditional random fields. Deep learning is used for improving the system robustness. The different modalities of brain MRI is considered such as T1, T2, flair constant of BRATS 2013, BRATS 2015. BRATS 2016 segmentation challenge. The analysis of previous literature and their techniques performance are shown in Table 1.

Table 1 Analysis of related paper

Author	Segmentation technique	Classifier	Performance	Dataset
Kumar et al. [49]	Gradient vector flow—boundary based technique	PCA-ANN	95.37%	55 patient-T1 weighted MR
Lashkari et al. [50]	Histogram equalization morphological operation	MLP model-ANN	98%	210 case-T1 weighted, T2 weighted MRI
Wang et al. [51]	–	Cascaded CNN model anisotropic CNN model	–	BRATS 2017
Byale et al. [52]	K-mean segmentation, GMM segmentation	Artificial neural network	93.33%	60 sample MRI
Kharrat et al. [53]	–	GA + SVM	94.44%	83 sample images—T2 weighted
Sharma et al. [37, 46]	Global thresholding, Anisotropic diffusion filtering	DE + ANN	94.34%	T1 weighted MRI
Ortiz et al. [54]	SOM clustering algorithm	–	–	IBSR
Shanthi et al. [55]	Fuzzy c-mean algorithm	ANN	–	–
El Abbadi et al. [56]	Morphological operations	Probabilistic neural network	98%	65 MR image dataset
El-Dahshan et al. [57]	–	FP-ANN	97%	70 MR images

Deep learning consists of multiprocessing layers which can handle larger complex hierarchy of data [58]. Deep artificial neural network is applied in numerous medical visualization analysis as it shows outstanding performance efficiency in comparison to other manual or semi-automatic techniques [59]. There are various deep learning algorithms such as deep Boltzmann machine, Stack auto-encoders, Convolutional neural network, fine tuning deep models for target task [60] etc. The CNN architecture involves many layers of pooling, activation and classification. The different CNN architecture used and their performance are discussed in Table 2.

Table 2 CNN techniques and its analysis

Author	Data	CNN model	Performance
Kamnitsas et al. [61]	BRATS 2015	Patch-wise CNN	DSC (0.9) (complete)
Zhao et al. [62]	BRATS 2013	Patch-wise CNN	Accuracy (0.81) (overall)
Nie et al. [63]	Private data	Semantic-wise CNN	DSC 85.5%(CSF)
Li et al. [64]	ILD	Single convolution layer	Accuracy (0.85) (overall)
Chao et al. [65]	MNIST	CaRENets	Accuracy (0.925) (overall)

3 Proposed Methodology

The steps for proposed study and contribution for development of automatic brain tumor segmentation system is shown as follows.

(a) The 61 sample cases of T1-weighted brain magnetic resonance images are obtained from IBSR (Brain segmentation repository). The 40 MS-free data sample images taken from Institute of neurology and genetics, at Nicosia Cyprus [66] and Laboratory of eHealth at the University of Cyprus [67]. The obtained images are normalized so that segmentation could be applied efficiently and properly.

(b) The skull stripping is essential fundamental process desired for segmentation and analysis of brain tumor MRI [68]. It is considered as important pre-processing phase for removing the non-cerebral tissue and plays role in clinical research of neuroimage applications [69]. In the proposed method, the new hybrid approach of adaptive particle swarm optimization [70, 71] and OTSU [29] are contributed to paper for performing improved skull stripping of brain. The hybridization is providing the optimized threshold value which improves the efficiency and reliability of system. The process flow of skull stripping is described as follows:

(i) After normalization of dataset, the Gaussian filter is applied for smoothing of image and removing the noise. The mathematical formulation of gaussian filter can be shown as

$$G(a, b) = \frac{1}{2\pi\sigma^2} e^{-\frac{z}{2\pi\sigma^2}} \tag{1}$$

$$Z = a^2 + b^2 \tag{2}$$

a represents the distance on x axis and b represents the distance on y-axis from origin, σ represents the standard deviation of gaussian distribution.

(ii) The evaluation of adaptive thresholding with the usage of intensity histogram. Adaptive thresholding is dynamic process of obtaining threshold value which would be dependent upon the neighboring pixels intensity.

(iii) Calculate the threshold value through hybrid approach of OSTU + APSO. This is providing excellent results when compared to other methods. The modified particle swarm optimization introduces two additional adaptive parameters to improve convergence speed named as adaptive factor, perturbation factor. The fitness function value is obtained by applying metaheuristic algorithm.

(iv) Morphological operation performs functions in relevance to shape and size of image. The structuring element is taken as input for extracting the useful features and representation of data. The four elementary mathematical operations of morphology in image processing can be represented

$$\text{Erosion}: \ C \ominus D = \{B|(D)_B \subseteq C\} \tag{4}$$

$$\text{Dilation}: \ C \oplus D = \{B|(D)_B \cap C \neq \emptyset\} \tag{5}$$

$$\text{Opening}: \ C \ominus D = C \ominus D \oplus D \tag{6}$$

$$\text{Closing}: \ C \ominus D = C \oplus D \oplus D \tag{7}$$

(v) The skull stripping is performed by keeping the extracted mask on input image so that extra cerebral tissue can be eliminated and region of interest can be proceeded for further operations.

(c) The denoising popular technique for brain MRI known as Anisotropic diffusion filtering is performed to improve the quality of extracted brain MRI by intensify the contrast between the regions. It is type of enhancement method to balance the different noise levels in image.

(d) Feature extraction is the most crucial step for evaluation of various parameters of image. On the basis of extracted features, the performance predictions and calculation could be obtained. In the proposed work, the 19 statistical and texture related parameters are extracted so that efficient system could be developed. GLCM method is used for extracting the features. GLCM (grey level co-occurrence matrix) is statistical method considering the spatial relationship of pixels. The 19 statistical features obtained in proposed work are autocorrela-

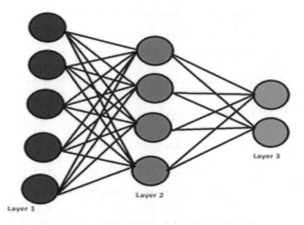

Fig. 2 Fully connected CNN

tion, cluster prominence, cluster shade, contrast, correlation, entropy, variance, dissimilarity, energy, difference Entropy, homogeneity, information measure of correlation 1, information measure of correlation 2, inverse difference, maximum probability, sum average, sum entropy, sum of square variance, sum variance. The extracted features are given as input to convolutional neural network so that tumor classification can be determined.

(e) Convolutional neural network is used as classification model for tumor detection system. The CNN outperforms in comparison to other classifiers if tremendous amount of data is needed to be handled. In the proposed work the three layers of convolutional neural network which uses activation function. The layer 1 and layer 2 uses [72] RELu activation function and layer 3 uses SoftMax activation function. The data is divided into 7:3 ratio. The layers are densely connected from one neuron to another neuron. The 98% of accuracy is achieved in proposed work. The diagrammatic representation is shown in Fig. 2.

The representation of procedural flow diagram of the proposed algorithm is represented in Fig. 3.

The proposed model algorithm steps are shown in table and are implemented on MATLAB R2018b

4 Experimental Results

4.1 Simulated Results

The proposed work uses 101 sample images of brain MRI. The 70% of data are used in training and 30% of data are used in testing. There are 61 sample case images of tumored IBSR dataset and 40 sample images of non-tumor MS-free brain

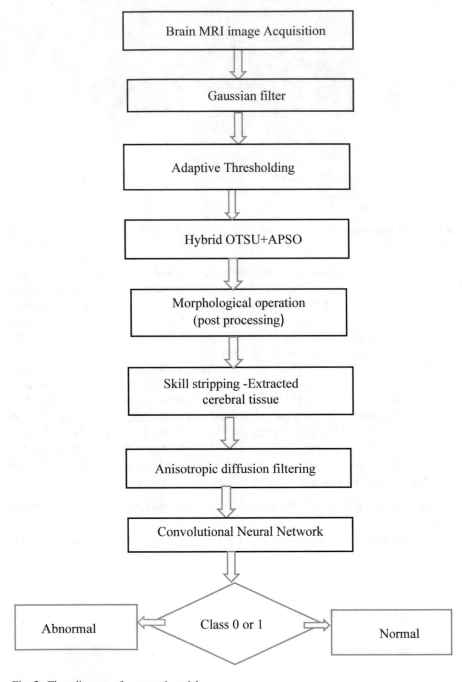

Fig. 3 Flow diagram of proposed model

MRI dataset. The adopted methodology performs hybridization of OTSU + APSO to determine optimal threshold value to obtain segmentation. The 19 features are extracted using GLCM (grey scale level matrix) which are used for training purpose in convolutional neural network. The 3 layers of convolutional neural network are densely connected to each other.

An example of IBSR tumored MRI are shown in figure (i) Normalize Image (ii) Gaussian Filter (iii) OTSU Thresholding + APSO (iv) and (v) Mathematical Morphology (vi) Skull Stripped Image (vii) Anisotropic Diffusion Filtering (viii) GLCM

An example of MS-free dataset non tumored MRI are shown in Fig. 4. (i) Normalize Image (ii) Gaussian Filter (iii) OTSU Thresholding + APSO (iv) and (v)

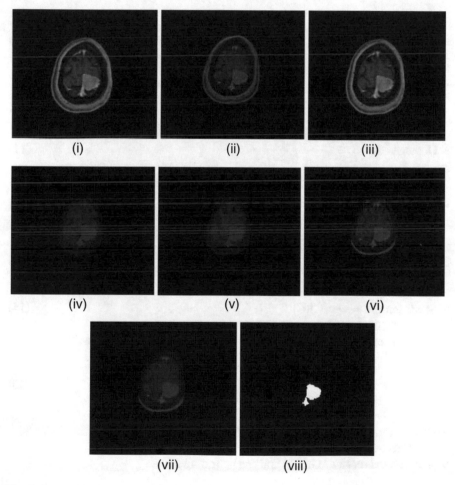

Fig. 4 Tumored images

Mathematical Morphology (vi) Skull Stripped Image (vii) Anisotropic Diffusion Filtering (viii) GLCM

The Table 4 represents the average fitness value and the segmented image quality metrics [73, 74] are calculated using peak signal to noise ratio (PSNR), structural similarity index matrix (SSIM) and Dice similarity parameter. The example outcomes of IBSR and MS-free dataset are shown from 1–25 to 26–40.

The mathematical formulation of few parameters of GLCM [75–77] technique are presented. All parameters are calculated for every sample MRI image of patients and the evaluation of parameters are independent of each other.

- Energy $= \sum_{i,j=0}^{L-1} (k_{i,j})^2$ \hfill (8)

- Entropy $= \sum_{i,j=0}^{L-1} -\lg(k_{i,j})k_{i,j}$ \hfill (9)

- Contrast $= \sum_{i,j=0}^{L-1} k_{i,j}(i-j)^2$ \hfill (10)

- Homogeneity $= \sum_{i,j=0}^{L-1} \frac{(k_{i,j})}{1+(i-j)^2}$ \hfill (11)

- Correlation $= \sum_{i,j=0}^{L-1} (k_{i,j}) \cdot \frac{(i-\mu)(j-\mu)}{(\sigma)^2}$ \hfill (12)

- Standard deviation $= \sqrt{\sum_{i,j=0}^{L-1} k_{i,j}(i-\mu)^2}$ \hfill (13)

- Shade feature $= \text{sgn}(C)|C|^{1/3}$ \hfill (14)

sgn represents the sign of real number which can be positive, negative

$$C = \sum_{i,j=0}^{L-1} \frac{Ki,j(i+j-2\mu)^3}{(\sigma)^3(\sqrt{2(1+C)})^3} \tag{15}$$

- $(k_{i,j}) =$ represents the attributes i, j of normalized symmetric matrix of GLCM
- L $=$ Number of levels
- μ represents mean of GLCM and can be formulated as

$$\mu = \sum_{i,j=0}^{L-1} \sum_{i,j=0}^{L-1} ik_{i,j} \tag{16}$$

- $(\sigma)^2$ represents the variance of all pixels.

$$(\sigma)^2 = \sum_{i,j=0}^{L-1} k_{i,j}(i - \mu)^2 \tag{17}$$

The training and testing data are evaluated on x axis and y axis in which x-axis represents the data with class label and y axis represents the data of encoded class labels. The three layers are considered having 38 neurons in 1st layer, 19 neurons in 2nd layer and class labels in 3rd layer. The categorial cross entropy loss function of deep learning is implemented. The values considered for the neural network measures are shown as follows:

learning rate $= 0.001$, $\beta_1 = 0.9$, $\beta_2 = 0.999$, activation function $=$ RELu and SoftMax

The accuracy of intelligent automatic brain tumor segmentation achieved is 98% and the loss function is 4% as presented in Table 7.

5 Results and Discussion

In the proposed study, Fig. 2 shows representation of fully connected neural network. The three layers are creating densely nested network which are aiming to provide classification results. The proceeding Fig. 3 presents the flow process of methodologies adopted and implemented. Table 3 contributes the proposed hybrid OTSU + APSO approach for obtaining the optimal or best threshold value for separating the cerebral tissue of brain and providing improved segmentation results. The Figs. 4 and 5 shows the simulated result of segmentation process. Further the Table 4 provides the mathematical measures of fitness value, PSNR, SSIM, Dice similarity. PSNR

Table 3 Hybrid algorithm of OTSU and APSO	Hybridization of OTSU + APSO for skull stripping and segmentation of MRI
	1. Normalization of acquired image dataset of brain MRI
	2. Skull stripping is applied to remove the cerebellum part and improving the segmentation efficiency
	3. Apply anisotropic diffusion for smoothing of image
	4. Feature extraction by using GLCM technique
	5. Training the convolutional neural network
	6. Testing the convolutional neural network to achieve the accuracy and measuring the performance

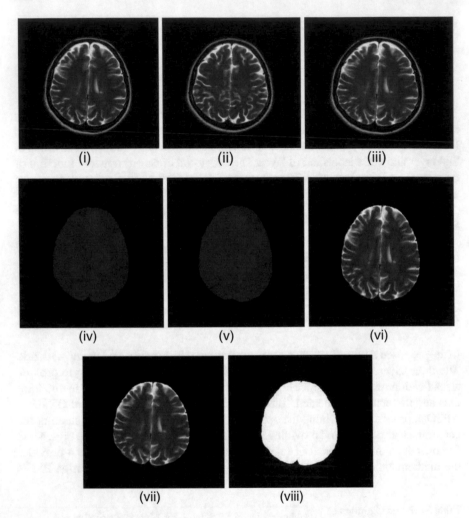

Fig. 5 Non tumored images

(peak signal to noise ratio) and SSIM determines the pixel difference and structural similarity quality metrics of segmented images. Dice similarity is statistic for evaluating the comparison of similarity between two images. The features extracted are

Table 4 Fitness values, PSNR, SSIM, Dice

Brain MRI	Fitness function value	PSNR	SSIM	Dice
Image 1	1.9913	45.7672	0.95441	0.98187
Image 2	1.9759	46.6009	0.95661	0.98443
Image 3	1.5481	46.8406	0.9566	0.98334
Image 4	1.9238	45.8822	0.9547	0.98435
Image 5	1.9751	45.9853	0.9554	0.98574
Image 6	1.9238	46.1083	0.9555	0.98439
Image 7	1.3843	46.2206	0.95593	0.98572
Image 8	1.8699	46.4003	0.95652	0.9855
Image 9	1.7917	46.6009	0.95661	0.98443
Image 10	1.0578	46.8304	0.9566	0.98334
Image 11	1.2111	45.7312	0.94363	0.9821
Image 12	1.9254	45.7589	0.94236	0.98205
Image 13	1.7899	45.8375	0.94208	0.98058
Image 14	1.7602	45.8722	0.94302	0.97993
Image 15	1.4607	45.8994	0.94247	0.98108
Image 16	1.6586	46.0584	0.941	0.98131
Image 17	1.586	46.1354	0.9399	0.97966
Image 18	1.6273	46.1821	0.93947	0.98021
Image 19	1.5251	46.2122	0.93956	0.98166
Image 20	1.897	46.3931	0.93939	0.98179
Image 21	1.1762	45.9448	0.98676	0.98639
Image 22	1.1409	45.7425	0.98918	0.98894
Image 23	1.8063	45.7003	0.98878	0.988
Image 24	1.9803	45.7184	0.98868	0.98867
Image 25	1.9913	45.7512	0.9879	0.98763
Image 26	1.6546	45.7907	0.98763	0.9875
Image 27	1.7807	45.8117	0.98747	0.98726
Image 28	1.8016	45.8636	0.98754	0.98687
Image 29	1.9627	45.4119	0.99608	1
Image 30	1.3021	45.6417	0.99591	1
Image 31	1.7973	46.1975	0.99614	1

(continued)

Table 4 (continued)

Brain MRI	Fitness function value	PSNR	SSIM	Dice
Image 32	1.4855	44.8914	0.99657	1
Image 33	1.8717	44.7648	0.99656	1
Image 34	1.794	44.7566	0.99655	1
Image 35	1.3735	44.7642	0.99655	1
Image 36	1.6309	44.8036	0.99653	1
Image 37	1.9422	44.8494	0.99649	1
Image 38	1.4554	44.9225	0.99642	1
Image 39	1.5642	45.0488	0.99633	1
Image 40	1.465	45.5681	0.99328	1
Image 41	1.8631	44.5622	0.99445	1
Image 42	1.4802	44.562	0.99443	1
Image 43	1.8498	44.5782	0.99441	1
Image 44	1.7799	44.6284	0.99433	1

provided as input for training the convolutional neural network. The Tables 5 and 6 represents the 19 features extracted from IBSR and MS-free dataset. The final parametric result of 101 sample images are shown in Table 7. It shows that proposed system achieves higher accuracy of 98%.

Table 5 Feature extraction of IBSR dataset

Sl. no.	Autocorrelation	Cluster prominence	Cluster shade	Contrast	Correlation	Difference entropy	Difference variance	Dissimilarity	Energy
1	1.0366	0.1692	0.0875	0.0039	0.8456	0.0256	0.0039	0.0039	0.971
2	1.0129	0.0649	0.0329	0.000613	0.93	0.0051	0.0006124	0.0006128	0.991
3	1.0211	0.106	0.0537	0.000766	0.9461	0.0063	0.0007653	0.0007659	0.985
4	1.0193	0.0946	0.0482	0.0013	0.9003	0.0101	0.0013	0.0013	0.986
5	1.0207	0.1016	0.0518	0.0013	0.9047	0.0103	0.0013	0.0013	0.985
6	1.0267	0.1285	0.0658	0.002	0.8893	0.0146	0.002	0.002	0.98
7	1.0273	0.1352	0.0688	0.0012	0.9367	0.009	0.0012	0.0012	0.98
8	1.1426	0.5984	0.3162	0.0066	0.9282	0.04	0.0066	0.0066	0.901
9	1.0612	0.2773	0.1441	0.0055	0.8692	0.0339	0.0054	0.0055	0.953
10	1.0612	0.2773	0.1441	0.0055	0.8692	0.0339	0.0054	0.0055	0.953
11	1.0542	0.2518	0.13	0.004	0.8924	0.0258	0.0039	0.004	0.959
12	1.046	0.2131	0.1101	0.004	0.8715	0.0263	0.004	0.004	0.965
13	1.0448	0.2075	0.1073	0.0041	0.8665	0.0267	0.0041	0.0041	0.965
14	1.0531	0.2461	0.1272	0.0041	0.8855	0.0268	0.0041	0.0041	0.96
15	1.0766	0.3429	0.1786	0.006	0.8838	0.0367	0.006	0.006	0.942
16	1.0494	0.2312	0.1192	0.0036	0.8923	0.0239	0.0036	0.0036	0.963
17	1.0271	0.1353	0.0687	0.000919	0.9495	0.0073	0.0009183	0.0009191	0.981
18	1.0278	0.1375	0.07	0.0012	0.9347	0.0094	0.0012	0.0012	0.98
19	1.0247	0.12	0.0613	0.0017	0.9001	0.0124	0.0017	0.0017	0.982
20	1.0284	0.1385	0.0707	0.0017	0.9141	0.0122	0.0017	0.0017	0.979

(continued)

Table 5 (continued)

Sl. no.	Entropy	Homogeneity	Information measure of correlation 1	Information measure of correlation 2	Inverse difference	Maximum probability	Sum average	Sum entropy	Sum of squares variance	Sum variance
1	0.0884	0.998	− 0.7144	0.306	0.998	0.9852	2.0257	0.0857	0.0127	0.0469
2	0.0321	0.9997	− 0.8621	0.218	0.9997	0.9953	2.0088	0.0317	0.0044	0.0169
3	0.0474	0.9996	− 0.8849	0.2691	0.9996	0.9925	2.0143	0.0468	0.0071	0.0276
4	0.0476	0.9993	− 0.8091	0.2502	0.9993	0.9927	2.0133	0.0467	0.0066	0.0251
5	0.0501	0.9993	− 0.8152	0.2582	0.9993	0.9922	2.0142	0.0492	0.0071	0.0269
6	0.0636	0.999	− 0.787	0.2813	0.999	0.9898	2.0184	0.0622	0.0091	0.0335
7	0.0598	0.9994	− 0.8655	0.2952	0.9994	0.9901	2.0186	0.059	0.0092	0.0335
8	0.2287	0.9967	− 0.8239	0.5236	0.9967	0.948	2.0973	0.2241	0.0463	0.1784
9	0.13	0.9973	− 0.7389	0.3759	0.9973	0.976	2.0426	0.1262	0.0209	0.078
10	0.13	0.9973	- 0.7389	0.3759	0.9973	0.976	2.0426	0.1262	0.0209	0.078
11	0.1136	0.998	− 0.779	0.3673	0.998	0.9793	2.0374	0.1108	0.0184	0.0695
12	0.1026	0.998	− 0.7488	0.3399	0.998	0.982	2.032	0.0998	0.0157	0.0589
13	0.1013	0.9979	− 0.7416	0.3354	0.9979	0.9823	2.0313	0.0984	0.0154	0.0574
14	0.113	0.9979	− 0.7681	0.3625	0.9979	0.9795	2.0368	0.1101	0.0181	0.0681
15	0.1522	0.997	− 0.7573	0.4115	0.997	0.9705	2.0531	0.148	0.0258	0.0973
16	0.1054	0.9982	− 0.7807	0.3553	0.9982	0.9811	2.0342	0.1029	0.0168	0.0635
17	0.058	0.9995	- 0.8886	0.2976	0.9995	0.9904	2.0184	0.0574	0.0091	0.0354
18	0.0609	0.9994	− 0.8616	0.2969	0.9994	0.9899	2.0189	0.0601	0.0094	0.0363
19	0.0585	0.9992	− 0.8052	0.2754	0.9992	0.9907	2.017	0.0574	0.0084	0.032
20	0.0643	0.9992	− 0.8261	0.2941	0.9992	0.9894	2.0195	0.0631	0.0096	0.0369

Table 6 Feature extraction of MS-free dataset

Sl. no.	Autocorrelation	Cluster prominence	Cluster shade	Contrast	Correlation	Difference entropy	Difference variance	Dissimilarity	Energy
1	1.8026	1.2847	0.724	0.003	0.9925	0.0202	0.003	0.003	0.605
2	1.7628	1.3004	0.7407	0.0029	0.9923	0.02	0.0029	0.0029	0.617
3	1.8009	1.2754	0.7183	0.007	0.9822	0.0416	0.0069	0.007	0.601
4	1.692	1.29	0.742	0.0134	0.9624	0.0712	0.0133	0.0134	0.629
5	1.8669	1.2537	0.6887	0.0032	0.9923	0.0215	0.0032	0.0032	0.585
6	1.7026	1.2846	0.7378	0.0148	0.9591	0.077	0.0146	0.0148	0.624
7	1.8477	1.2633	0.7002	0.0032	0.9921	0.0216	0.0032	0.0032	0.591
8	1.8308	1.272	0.7099	0.003	0.9926	0.0203	0.003	0.003	0.596
9	1.4108	1.1807	0.6667	0.0111	0.9535	0.0611	0.011	0.0111	0.75
10	1.5036	1.2878	0.7367	0.0032	0.9887	0.0214	0.0032	0.0032	0.717
11	1.5261	1.2755	0.7326	0.0105	0.9639	0.0584	0.0104	0.0105	0.698
12	1.6836	1.2946	0.7453	0.0122	0.9655	0.0659	0.0121	0.0122	0.634
13	1.9701	1.1907	0.6095	0.006	0.9863	0.0366	0.006	0.006	0.556
14	2.0582	1.1464	0.5345	0.0025	0.9945	0.0174	0.0025	0.0025	0.541
15	2.03	1.163	0.5618	0.0024	0.9947	0.0168	0.0024	0.0024	0.547
16	1.9922	1.1852	0.5961	0.0023	0.9948	0.0163	0.0023	0.0023	0.555
17	1.9654	1.2008	0.6188	0.0023	0.9946	0.0166	0.0023	0.0023	0.561
18	1.9229	1.2255	0.6522	0.0022	0.9948	0.0159	0.0022	0.0022	0.572
19	1.8499	1.2643	0.7003	0.0023	0.9942	0.0165	0.0023	0.0023	0.591
20	1.6643	1.3222	0.7628	0.0033	0.9905	0.0221	0.0033	0.0033	0.651

(continued)

Table 6 (continued)

Sl. no.	Entropy	Homogeneity	Information measure of correlation 1	Information measure of correlation 2	Inverse difference	Maximum probability	Sum average	Sum entropy	Sum of squares variance	Sum variance
1	0.6011	0.9985	−0.9659	0.8214	0.9985	0.7305	2.536	0.5991	0.1962	0.0469
2	0.5871	0.9985	−0.9654	0.8159	0.9985	0.7438	2.5095	0.5851	0.1898	0.0169
3	0.6221	0.9965	−0.9299	0.8129	0.9965	0.7284	2.5363	0.6173	0.1962	0.0276
4	0.6117	0.9933	−0.873	0.7825	0.9933	0.7604	2.4658	0.6024	0.1787	0.0251
5	0.6229	0.9984	−0.9648	0.8288	0.9984	0.7089	2.579	0.6207	0.2057	0.0269
6	0.6217	0.9926	−0.8638	0.782	0.9926	0.756	2.4733	0.6115	0.1807	0.0335
7	0.6171	0.9984	−0.9644	0.8265	0.9984	0.7153	2.5662	0.6149	0.203	0.0335
8	0.6105	0.9985	−0.9662	0.8249	0.9985	0.7211	2.5549	0.6084	0.2005	0.1784
9	0.4596	0.9944	−0.8587	0.7066	0.9944	0.8557	2.2775	0.4519	0.1195	0.078
10	0.4738	0.9984	−0.9549	0.7611	0.9984	0.83	2.3368	0.4716	0.14	0.078
11	0.5225	0.9947	−0.8811	0.7489	0.9947	0.8176	2.3542	0.5152	0.1458	0.0695
12	0.6029	0.9939	−0.8817	0.7833	0.9939	0.764	2.4598	0.5945	0.177	0.0589
13	0.6664	0.997	−0.9425	0.8337	0.997	0.6726	2.6487	0.6622	0.2192	0.0574
14	0.6667	0.9988	−0.9733	0.8471	0.9988	0.6456	2.7063	0.665	0.2284	0.0681
15	0.6601	0.9988	−0.9741	0.8453	0.9988	0.6551	2.6874	0.6585	0.2256	0.0973
16	0.6512	0.9988	−0.9745	0.8426	0.9988	0.6677	2.6622	0.6496	0.2215	0.0635
17	0.6449	0.9988	−0.9739	0.8402	0.9988	0.6766	2.6444	0.6433	0.2184	0.0354
18	0.6332	0.9989	−0.9746	0.8366	0.9989	0.6909	2.616	0.6316	0.2131	0.0363
19	0.6127	0.9988	−0.9727	0.8286	0.9988	0.7151	2.5674	0.611	0.2032	0.032
20	0.5509	0.9984	−0.9594	0.7987	0.9984	0.7764	2.4439	0.5486	0.1727	0.0369

Table 7 Final Result

Classification model	Accuracy (%)	Loss function (%)
Convolutional neural network	98	4

6 Conclusion

The intelligent computer aided diagnosis system for brain tumor segmentation is proposed in this research work. The data of 61 IBSR tumored magnetic resonance imagining and 40 MR-free non-tumored data is considered and observed. The fusion of OTSU embedded adaptive particle swarm optimization is used for obtaining the best threshold value to be applied for segmentation. The automatic brain tumor segmentation is performed involving skull stripping technique. The 19 statistical and texture features are extracted using GLCM which are used in training of convolutional neural network. The categorial cross entropy loss function, RELu and SoftMax activation function is taken in convolutional neural network. Convolutional neural network contributed in improving the classification results of adopted methodology by achieving the accuracy of 98% which is better than other existing system. In future the model will be designed for different modalities of data and other metaheuristic algorithm can be used for improving the performance of the diagnosed system.

Acknowledgements Authors would like to thank for the support and valuable time provided by Amity university, Noida

References

1. Roy S, Bandyopadhyay SK (2012) Detection and quantification of brain tumor from MRI of brain and it's symmetric analysis. Int J Inf Commun Technol Res 2(6)
2. McAuliffe MJ, Lalonde FM, McGarry D, Gandler W, Csaky K, Trus BL (2001) Medical image processing, analysis and visualization in clinical research. In: Proceedings 14th IEEE symposium on computer-based medical systems. CBMS 2001. IEEE, pp 381–386
3. Despotović I, Goossens B, Philips W (2015) MRI segmentation of the human brain: challenges, methods, and applications. Comput Math Methods Med 2015
4. Zaitoun NM, Aqel MJ (2015) Survey on image segmentation techniques. Proc Comput Sci 65:797–806
5. Mirjalili S, Lewis A (2016) The whale optimization algorithm. Adv Eng Softw 95:51–67
6. Tang YG, Liu D, Guan XP (2007) Fast image segmentation based on particle swarm optimization and two-dimension Otsu method. Control Decis 22(2):202
7. Li L, Sun L, Guo J, Han C, Zhou J, Li S (2017) A quick artificial bee colony algorithm for image thresholding. Information 8(1):16
8. Samantaa S, Dey N, Das P, Acharjee S, Chaudhuri SS (2013) Multilevel threshold based gray scale image segmentation using cuckoo search. arXiv:1307.0277
9. Soffer S, Ben-Cohen A, Shimon O, Amitai MM, Greenspan H, Klang E (2019) Convolutional neural networks for radiologic images: a radiologist's guide. Radiology, 290(3):180547

10. Li Q, Cai W, Wang X, Zhou Y, Feng DD, Chen M (2014). Medical image classification with convolutional neural network. In: 2014 13th international conference on control automation robotics & vision (ICARCV), December 2014. IEEE, pp 844–848
11. Schmidhuber J (2015) Deep learning in neural networks: an overview. Neural Netw 61:85–117
12. Suzuki K (2017) Overview of deep learning in medical imaging. Radiol Phys Technol 10(3):257–273
13. Erickson BJ, Korfiatis P, Akkus Z, Kline TL (2017) Machine learning for medical imaging. Radiographics 37(2):505–515
14. Sutskever I, Vinyals O, Le QV (2014) Sequence to sequence learning with neural networks. In: Advances in neural information processing systems, pp 3104–3112
15. Tajbakhsh N, Shin JY, Gurudu SR, Hurst RT, Kendall CB, Gotway MB, Liang J (2016) Convolutional neural networks for medical image analysis: full training or fine tuning? IEEE Trans Med Imaging 35(5):1299–1312
16. Lee JG, Jun S, Cho YW, Lee H, Kim GB, Seo JB, Kim N (2017) Deep learning in medical imaging: general overview. Korean J Radiol 18(4):570–584
17. Lo SCB, Chan HP, Lin JS, Li H, Freedman MT, Mun SK (1995) Artificial convolution neural network for medical image pattern recognition. Neural Netw 8(7–8):1201–1214
18. Litjens G, Kooi T, Bejnordi BE, Setio AAA, Ciompi F, Ghafoorian M, Sánchez CI (2017) A survey on deep learning in medical image analysis. Med Image Anal 42:60–88
19. LeCun Y, Kavukcuoglu K, Farabet C (2010) Convolutional networks and applications in vision. In: Proceedings of 2010 IEEE international symposium on circuits and systems, May 2010. IEEE, pp 253–256
20. Talo M, Baloglu UB, Yıldırım Ö, Acharya UR (2019) Application of deep transfer learning for automated brain abnormality classification using MR images. Cogn Syst Res 54:176–188
21. Balafar MA, Ramli AR, Saripan MI, Mashohor S (2010) Review of brain MRI image segmentation methods. Artif Intell Rev 33(3):261–274
22. Villanueva-Meyer JE, Chang P, Lupo JM, Hess CP, Flanders AE, Kohli M (2019) Machine learning in neurooncology imaging: from study request to diagnosis and treatment. Am J Roentgenol 212(1):52–56
23. Kheirollahi M, Dashti S, Khalaj Z, Nazemroaia F, Mahzouni P (2015) Brain tumors: special characters for research and banking. Adv Biomed Res 4
24. Bauer S, Wiest R, Nolte LP, Reyes M (2013) A survey of MRI-based medical image analysis for brain tumor studies. Phys Med Biol 58(13):R97
25. Drevelegas A, Papanikolaou N (2011) Imaging modalities in brain tumors. In Imaging of brain tumors with histological correlations. Springer, Berlin, pp 13–33
26. Roslan R, Jamil N, Mahmud R (2011) Skull stripping magnetic resonance images brain images: region growing versus mathematical morphology. Int J Comput Inf Syst Ind Manag Appl 3:150–158
27. Ségonne F, Dale AM, Busa E, Glessner M, Salat D, Hahn HK, Fischl B (2004) A hybrid approach to the skull stripping problem in MRI. Neuroimage 22(3):1060–1075
28. Patel J, Doshi K (2014) A study of segmentation methods for detection of tumor in brain MRI. Adv Electron Electr Eng 4(3):279–284
29. Park JG, Lee C (2009) Skull stripping based on region growing for magnetic resonance brain images. NeuroImage 47(4):1394–1407
30. Ahmmed R, Swakshar AS, Hossain MF, Rafiq MA (2017). Classification of tumors and it stages in brain MRI using support vector machine and artificial neural network. In: 2017 International conference on electrical, computer and communication engineering (ECCE), February. IEEE, pp 229–234
31. Cabria I, Gondra I (2017) MRI segmentation fusion for brain tumor detection. Inf Fusion 36:1–9
32. Ayachi R, Amor NB (2009). Brain tumor segmentation using support vector machines. In: European conference on symbolic and quantitative approaches to reasoning and uncertainty, July 2009. Springer, Berlin, pp 736–747

33. Soleimani V, Vincheh FH (2013). Improving ant colony optimization for brain MRI image segmentation and brain tumor diagnosis. In: 2013 first Iranian conference on pattern recognition and image analysis (PRIA), March 2013. IEEE, pp 1–6
34. Jothi G (2016) Hybrid Tolerance Rough Set-Firefly based supervised feature selection for MRI brain tumor image classification. Appl Soft Comput 46:639–651
35. Manic KS, Priya RK, Rajinikanth V (2016) Image multithresholding based on Kapur/Tsallis entropy and firefly algorithm. Indian J Sci Technol 9(12):89949
36. Sharma M, Purohit GN, Mukherjee S (2018) Information retrieves from brain MRI images for tumor detection using hybrid technique K-means and artificial neural network (KMANN). In: Networking communication and data knowledge engineering. Springer, Singapore, pp 145–157
37. Jafari M, Shafaghi R (2012) A hybrid approach for automatic tumor detection of brain MRI using support vector machine and genetic algorithm. Glob J Sci engineering and Technol 3:1–8
38. Jiang J, Trundle P, Ren J (2010) Medical image analysis with artificial neural networks. Comput Med Imaging Graph 34(8):617–631
39. Goodband JH, Haas OCL, Mills JA (2008) A comparison of neural network approaches for on-line prediction in IGRT. Med Phys 35(3):1113–1122
40. Lo SC, Li H, Freedman MT (2003) Optimization of wavelet decomposition for image compression and feature preservation. IEEE Trans Med Imaging 22(9):1141–1151
41. Suzuki K, Horiba I, Sugie N (2003) Neural edge enhancer for supervised edge enhancement from noisy images. IEEE Trans Pattern Anal Mach Intell 25(12):1582–1596
42. Havaei M, Davy A, Warde-Farley D, Biard A, Courville A, Bengio Y, Larochelle H (2017) Brain tumor segmentation with deep neural networks. Med Image Anal 35:18–31
43. Gao XW, Hui R, Tian Z (2017) Classification of CT brain images based on deep learning networks. Comput Methods Progr Biomed 138:49–56
44. Pereira S, Pinto A, Alves V, Silva CA (2016) Brain tumor segmentation using convolutional neural networks in MRI images. IEEE Trans Med Imaging 35(5):1240–1251
45. Sharma A, Kumar S, Singh SN (2018) Brain tumor segmentation using DE embedded OTSU method and neural network. Multidimens Syst Signal Process 30(3):1263–1291
46. Mohan G, Subashini MM (2018) MRI based medical image analysis: survey on brain tumor grade classification. Biomed Signal Process Control 39:139–161
47. Chen H, Dou Q, Yu L, Qin J, Heng PA (2018) VoxResNet: deep voxelwise residual networks for brain segmentation from 3D MR images. NeuroImage 170:446–455
48. Zhao X, Wu Y, Song G, Li Z, Zhang Y, Fan Y (2018) A deep learning model integrating FCNNs and CRFs for brain tumor segmentation. Med Image Anal 43:98–111
49. Kumar V, Sachdeva J, Gupta I, Khandelwal N, Ahuja CK (2011) Classification of brain tumors using PCA-ANN. In: 2011 world congress on information and communication technologies, December 2011. IEEE, pp 1079–1083
50. Lashkari A (2010) A neural network based method for brain abnormality detection in MR images using Gabor wavelets. Int J Comput Appl 4(7):9–15
51. Wang G, Li W, Ourselin S, Vercauteren T (2017) Automatic brain tumor segmentation using cascaded anisotropic convolutional neural networks. In: International MICCAI brainlesion workshop, September 2017. Springer, Cham, pp 178–190
52. Byale H, Lingaraju GM, Sivasubramanian S (2018) Automatic segmentation and classification of brain tumor using machine learning techniques. Int J Appl Eng Res 13(14):11686–11692
53. Kharrat A, Gasmi K, Messaoud MB, Benamrane N, Abid M (2010) A hybrid approach for automatic classification of brain MRI using genetic algorithm and support vector machine. Leonardo J Sci 17(1):71–82
54. Ortiz A, Górriz JM, Ramírez J, Salas-Gonzalez D, Llamas-Elvira JM (2013) Two fully-unsupervised methods for MR brain image segmentation using SOM-based strategies. Appl Soft Comput 13(5):2668–2682
55. Shanthi KJ, Sasikumar MN, Kesavadas C (2010) Neuro-fuzzy approach toward segmentation of brain MRI based on intensity and spatial distribution. J Med Imaging Radiat Sci 41(2):66–71

56. El Abbadi NK, Kadhim NE (2017) Brain cancer classification based on features and artificial neural network. Brain 6(1)
57. El-Dahshan ESA, Hosny T, Salem ABM (2010) Hybrid intelligent techniques for MRI brain images classification. Digit Signal Process 20(2):433–441
58. Gibson E, Li W, Sudre C, Fidon L, Shakir DI, Wang G, Whyntie T (2018) NiftyNet: a deep-learning platform for medical imaging. Comput Methods Progr Biomed 158:113–122
59. Akkus Z, Galimzianova A, Hoogi A, Rubin DL, Erickson BJ (2017) Deep learning for brain MRI segmentation: state of the art and future directions. J Digit Imaging 30(4):449–459
60. Shen D, Wu G, Suk HI (2017) Deep learning in medical image analysis. Annu Rev Biomed Eng 19:221–248
61. Kamnitsas K, Ledig C, Newcombe VF, Simpson JP, Kane AD, Menon DK, Glocker B (2017) Efficient multi-scale 3D CNN with fully connected CRF for accurate brain lesion segmentation. Med Image Anal 36:61–78
62. Zhao L, Jia K (2015) Deep feature learning with discrimination mechanism for brain tumor segmentation and diagnosis. In: 2015 international conference on intelligent information hiding and multimedia signal processing (IIH-MSP), September 2015. IEEE, pp 306–309
63. Nie D, Wang L, Gao Y, Sken D (2016). Fully convolutional networks for multi-modality isointense infant brain image segmentation. In: 2016 IEEE 13th international symposium on biomedical imaging (ISBI), April 2016. IEEE, pp 1342–1345
64. Li Q et al (2014) Medical image classification with convolutional neural network. In: 2014 13th international conference on control automation robotics & vision (ICARCV). IEEE
65. Chao J et al (2019) CaRENets: compact and resource-efficient CNN for homomorphic inference on encrypted medical images. arXiv:1901.10074
66. Loizou CP, Petroudi S, Seimenis I, Pantziaris M, Pattichis CS (2015) Quantitative texture analysis of brain white matter lesions derived from T2-weighted MR images in MS patients with clinically isolated syndrome. J Neuroradiol 42(2):99–114
67. http://www.medinfo.cs.ucy.ac.cy/
68. Zhuang AH, Valentino DJ, Toga AW (2006) Skull-stripping magnetic resonance brain images using a model-based level set. NeuroImage 32(1):79–92
69. Zhan ZH, Zhang J, Li Y, Chung HSH (2009) Adaptive particle swarm optimization. IEEE Trans Syst Man Cybern Part B (Cybern) 39(6):1362–1381
70. Duraisamy SP, Kayalvizhi R (2010) A new multilevel thresholding method using swarm intelligence algorithm for image segmentation. J Intell Learn Syst Appl 2(03):126
71. Sezgin M, Sankur B (2004) Survey over image thresholding techniques and quantitative performance evaluation. J Electron Imaging 13(1):146–166
72. Ker J, Wang L, Rao J, Lim T (2018) Deep learning applications in medical image analysis. IEEE Access 6:9375–9389
73. Hore A, Ziou D (2010). Image quality metrics: PSNR vs. SSIM. In: 2010 20th international conference on pattern recognition, August 2010. IEEE, pp 2366–2369
74. Yeghiazaryan V, Voiculescu I (2015) An overview of current evaluation methods used in medical image segmentation. Technical report CS-RR-15-08 Department of Computer Science, University of Oxford, Oxford, UK
75. Jain A, Zongker D (1997) Feature selection: Evaluation, application, and small sample performance. IEEE Trans Pattern Anal Mach Intell 19(2):153–158
76. Malegori C, Franzetti L, Guidetti R, Casiraghi E, Rossi R (2016) GLCM, an image analysis technique for early detection of biofilm. J Food Eng 185:48–55
77. Zayed N, Elnemr HA (2015) Statistical analysis of haralick texture features to discriminate lung abnormalities. J Biomed Imaging 2015:12

Analysis and Visualization of User Navigations on Web

Honey Jindal, Neetu Sardana and Raghav Mehta

Abstract The web is the largest repository of data. The user frequently navigates on the web to access the information. These navigational patterns are stored in weblogs which are growing exponentially with time. This increase in voluminous weblog data raises major challenges concerning handling big data, understanding navigation patterns and the structural complexity of the web, etc. Visualization is a process to view the complex large web data graphically to address these challenges. This chapter describes the various aspects of visualization with which the novel insights can be drawn in the area of web navigation mining. To analyze user navigations, visualization can be applied in two stages: post pre-processing and post pattern discovery. First stage analyses the website structure, website evolution, user navigation behaviour, frequent and rare patterns and detecting noise. Second stage analyses the interesting patterns obtained from prediction modelling of web data. The chapter also highlights popular visualization tools to analyze weblog data.

Keywords Navigation · Visualization · Pattern · Website · User · Weblogs · Analysis

1 Introduction

In the last 20 years, the web has become the largest source of information. The web is expanding exponentially every year. This increase in the web has raised a lot of challenges like handling the large volume of data, handling the structural complexity of web sites, understanding user navigations, etc.

H. Jindal (✉) · N. Sardana (✉) · R. Mehta (✉)
Jaypee Institute of Information and Technology, Noida, India
e-mail: honey.cs0990@gmail.com

N. Sardana
e-mail: neetu.sardana@jiit.com

R. Mehta
e-mail: raghav.mehta.17@gmail.com

© Springer Nature Switzerland AG 2020
J. Hemanth et al. (eds.), *Data Visualization and Knowledge Engineering*,
Lecture Notes on Data Engineering and Communications Technologies 32,
https://doi.org/10.1007/978-3-030-25797-2_9

Users navigate on the web to access the information of their interest. These navigations patterns are stored in web log files. These files can help in extracting useful hidden facts. Initially, during the pre-processing stage, weblogs are cleaned, in which irrelevant information is removed, and noise is filtered. Post this stage the weblogs can be used to discover interesting and useful patterns by applying supervised or unsupervised learning techniques. Data present in the weblogs can be analyzed in various dimensions by visualizing it graphically. Visualization is a process to present information in the varied visual forms, i.e., graphs, trees, maps, charts, etc. This would help people to understand the insights of the large volume of data. Patterns, correlations, and trends which were undetected can be discovered using data visualization. Discovering web navigational patterns, trend and analyzing their results is undeniably gives advantages to web designers and website owners.

The analysis of web navigation patterns would steer ample web application like website design [1], business intelligence modelling [2, 3], promotional advertisement and personalization [4], and e-commerce [5].

Visualization of user navigation patterns is an essential prerequisite for generating effective prediction system. Visualization is an important part of pre-processing, pattern discovery and analyzing. It may be a part of the exploratory process [6]. Patterns can be visualized at two stages. Initially, data can be visualized once data is pre-processed and finally we can visualize patterns after data modelling. Both stages give insights into data with different perceptive. For example, to identify outliers, to discover new trends or patterns, check the quality of patterns, an efficiency of the model and evaluate the strength of evidence. The overall framework of the visualization of web navigation is shown in Fig. 1. The figure shows all the components of Web Navigation Mining and Visualization. The components are web log files, Data Cleaning and Pre-processing, Pattern discovery and stage A & B data visualizations.

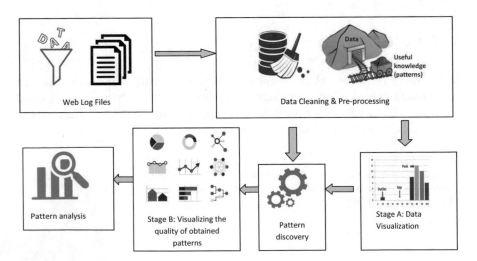

Fig. 1 Overall framework of web navigation mining

The following section describes the components of the given framework in a detailed manner. It also explains which questions can be addressed in two identified stages of visualizations.

2 Framework for Web Navigation Mining

2.1 Web Log File

Weblog file [7] is the record of user(s) activities performed during website navigation. The web server automatically creates this log file. Each hit to the website, including web page view, an image is logged in this file. It consists of information about users visiting the site, the source from which they came, and their objective to surf the web site. For every request, raw web log file store one line. A line is consist of various tokens separated by spaces. Hyphen (-) is placed when the token has no value. A line in a log file might look like the following:

> 192.168.1.3 - - [18/Feb/2000:13:33:37 -0600] "GET /HTTP/1.0" 200 5073

Some common log file types supported by the web server:

(a) **Common Log Format (Access)** [8]: This format is also known as the NCSA Common log format. The W3C working group defines the format. It is a standard text file format used for generating server log files. The Common Log Format has the following syntax:

> "%h %l %u %t \"%r\" %>s %b"

where h is host id, l is client identity, u is user id, t is time and date, r is request, s is status and b is bytes
For example [9],

> 27.0.0.1 user-identifier zoya [11/Aug/2013:10:50:34 -0700] "GET /apache_pb.gif HTTP/1.0" 400 2326

Where 27.0.0.1 is the IP address of the user, user-identifier is the identity of the user, zoya is the user id, 11/Aug/2013:10:50:34 -0700 denotes date and time of the request, GET /apache_pb.gif HTTP/1.0 is the http request method, 400 is the http status code, 400 is client error, 2326 is the size of the object returned to the client which is measured in bytes.

(b) **Extended (Access, Referrer, and Agent)** [8]: This format has two types: NCSA (National Centre for Supercomputing Applications) and the W3C (World Wide Web Consortium). The NCSA extended log format is the common log format appended with the agent and referrer information.

- **NCSA's extended format** is defined by the following string:

> "%h %l %u %t \"%r\" %>s %b \"%{Referrer}i\"\"%{User-agent}i\"

- **W3C extended log format** [10]: This log file is handled by Internet Information Server (IIS). W3C is a extended and customizable ASCII format with several fields like web browser, time, user IP address, method, URI Stem, HTTP Status, and HTTP Version. The time used here is Greenwich Mean Time (GMT).

#Software: Microsoft IIS 5.0 #Version: 2.0 #Date: 2004-06-05 20:40:12
20:40:12 183.18.255.255 GET /default.html 200 HTTP/2.0.

3. IIS (Internet Information Service) [10]: This log file format is simple. It consist several fields like time of requesting the web file, user IP address, request command, requested file, and response status code.

> 02:40:12 128.0.0.1 GET/404
> 02:50:35 128.0.0.1 GET/homepage.html 200
> 02:51:10 128.0.0.1 GET/graphs/prediction.gif 200
> 03:10:56 128.0.0.1 GET/search.php 304
> 03:45:05 128.0.0.1 GET/admin/style.css 200
> 04:10:32 128.0.0.1 GET/teachers.ico 404.

2.2 Pre-processing

The weblog contains large information of user navigation history. Sometimes, all information is not meaningful. For example, a user while browsing a web page, downloads images, videos, JS and CSS files. Therefore, we need to filter out this irrelevant information from the weblog file. The objective of the weblog file pre-processing is to remove irrelevant data from the web log file and reducing the amount of data. This process includes three steps: Data Cleaning, Users Identification, Session Identification and Session Reconstruction.

(a) **Data cleaning**: There are three ways to filter the data:

URL or website. The user session consists of information of user HTML navigated web page. The suffix for a gif, jpg, js files has no significance in session formation; thus it can be removed from the log file. For some special sites which generate content dynamically, filtering rules must be adjusted according to the requirement.

Request action and return status. The GET request actions are retained. Also, successful browsing records are retained while browsing record with error code is removed. Sometimes, the error logs of the website is useful for the administrator to analyze security.

Request IP: To get rid of the access from the automated requester, a robot IP filter list can be created.

(b) **User Identification**

The user refers to an individual accessing server through a web browser [11]. A weblog file can distinguish the user's through user IP, user agents, and session cookies. This heuristic assumes that if two weblog entries are having the same IP address but different user agents, then these entries may belong to two different users. Figure 2 illustrated four user and their navigation paths. These users are identified through their IP address. Here, P_i represents the web page.

(c) **Session Identification**

A session identification is a technique which records user navigations during one visit into a session. In web navigation log files, sessions are defined as the sequence of web pages traversed by the users. The session identifier stops working when the users' process terminates, when the internet connection is lost or when a timeout. The session identification or construction techniques use three fields when processing web server logs: the IP address of the user, request time of the web page and requested URL address. Session construction techniques may differ because some use time information while other use navigation information of the user. The session identification heuristics are categorised into four; time-oriented, navigation oriented, integer programming. Based on session construction heuristics the outcome of this

Fig. 2 Pages navigated by the users

| User 1: P3, P6, P9, P17, P13, P7, P40 |
| User 2: P3, P7, P8, P10, P35, P4 |
| User 3: P6, P10, P24, P21 |
| User 4: P8, P19, P5, P4 |

Fig. 3 Session identification
from weblogs

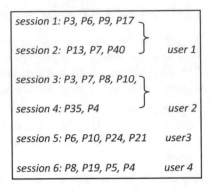

phase is presented in Fig. 3 which generates sessions from Fig. 2. Herein six sessions
are formed corresponding to four users.

Time-Oriented heuristics: Time-oriented heuristics have used the time-based threshold to create sessions: session time and page-stay. Catedge and Pitkow [12] measured mean time spend in a website is 9.3 min. Later, he derived a new duration time
25.5 min by adding 1,5 standard deviations. This has been rounded to 30 min. This
becomes the thumb rule for many applications [13, 14] and used widely as session
duration cut-off. During navigation with session duration 30 min, if a time between
one accessed page is longer than session duration cut-off, then the next request web-
page will be considered as a new session. This observation results in another type of
heuristic known as page stay time. Generally, this threshold is set to 10 min according
to [12]. If the difference between the timestamp of current navigated web pages and
next navigation web page is greater than the specified threshold, then the current session is terminated, and the next session will start. The page stay-time [15] is affected
by the page content, the time needed to load the page components and the transfer
speed of communication line.

Time-oriented heuristics do not follow link information of the website, i.e., some
useful information of the navigation paths are missed. Moreover, the order of web
pages navigated by the users' may not be recorded correctly in the time based heuristics due to proxies or browser cache.

Navigation-Oriented: This heuristic does not consider time-based thresholds; rather
it uses graph-based modelling. In navigation-oriented, web pages are considered
as nodes and hyperlinks are considered as the directed edges between the nodes.
According to Cooley [14], a requested web page P_i which is not reachable from the
navigated web pages should be assigned to the new session. This heuristic is taken
into account that P_i need not to accessible from the visited web pages; rather a user
may backtrack to visited pages to reach P_i. This backward movement may not be
recorded in the log file due to user cache. In this case, the heuristic discovers the
shortest subsequence leading to P_i and add it to the user session.

The sub-sessions generated from clicking the back button do not imply strong
correlations among web pages [16]. Therefore, it becomes a major issue. Cooley
et al. [13, 14] show that backward navigation occurs due to the location of the web

page rather than their content. The occurrence of this backward navigation produces noise in the session. Another drawback of this method is the increase in session length. Addition of backward navigations in the session results into longer patterns which will further arise the network complexity and become computationally more expensive.

Integer Programming Approach: Dell et al. [17, 18] proposed the integer pro gram-ming approach which partitioned sessions into chunks using the IP address and agent information. To obtain longer session, these chunks are divided into sessions using logarithm function. The objective of this logarithm function is to assign web pages in a chunk to different web session. This implies that the reconstructed session have unique web pages. In an improved version [19], a backward pattern is added which allows repetition of web pages at Kth and $(K + 2)$th position. However, this technique does not cover all combination of user return to the page like $(k - 3)$th, etc. This addition also increases noise in the session and page repetition.

(d) Session Reconstruction

Session reconstruction is essential to provide correct assignments of navigations to the prediction model. The performance of web navigation prediction models relies on the quality of session inputs [16, 20]. There are some methods to regenerate sessions: Dynamic time-oriented heuristic [21], Navigation Tree [5, 22], Smart-SRA [16], BBCom [20], BBDcom [20]. Dynamic time-oriented heuristic [21] uses session duration and page-stay time thresholds. This function generates more effective site-specific thresholds and replaces either session duration or page stay threshold during session construction. Navigation Tree is formed using maximal forward length sequences. The sessions are generated from the root node to leaf node traversals. The sub-sessions obtained from this technique may consist of repeated web pages which require high storage and computational cost. Another technique called Smart-SRA (Smart Session Reconstruction Algorithm) [16] was proposed which takes session duration, page-stay-time, maximum length sessions, web topology collectively to eliminate backward browsing sequences. This technique produces correlated web sessions, but the process to reconstruct the session is quite long. To find the shortest path between the source (entry) web page and destination (exit) web page, Jindal et al. [20] proposed two backward browsing elimination techniques, Compressed back-ward browsing, BBcom and Decompressed backward browsing, BBDcom. BBcom compresses the navigation sequences and reduces the redundancy of the web pages. Whereas BBDcom decomposes the session into multiple sub sessions such that each sub-session consists web pages which were traversed in forward directions.

Once the session is generated, visualization tools are employed to check their effectiveness. Session 2.3 presents challenges of web navigation mining. It also describes how the visualization of data can help to provide solutions for the identified challenges.

2.3   Post Pre-processing Visualization: Stage 1

Once the data residing in weblogs is pre-processed, the visualization can be performed to know useful facts about the website. There are varied charts available to visualize the website with varied perspectives. The visualization can help to know the frequent and infrequent navigation patterns, depicting website structure from website navigations, explore the evolution of website with time, Identify entry and exit point from the navigational structure, Visualize the traffic and popularity of the Website(s) from Web Server logs, Classifying users based on navigations and noise detection to identify usual & unusual users.

(a)   *Identify Expected and Unexpected Frequent Patterns*

While analyzing the website navigation patterns, it's imperative to know the set of pages that are frequently accessed by the users as it signifies the popularity of the pages and also helps the website owners to make it richer in terms of content. Similarly, it's vital to know the rarely accessed set of pages as it helps in restructuring the website.

An interactive visual tool to analyze frequent patterns, called FpVAT [23] provides effective visual for data analysis. FpVAT consists of two modules: RdViz and FpViz. RdViz visualizes raw data, which helps the users to derive insight from the large raw data quickly. The second module, FpViz visualizes frequent patterns which are used to detect the number of frequent patterns expected and to discover the number of frequent patterns unexpected. Figure 4 presents the visualization of raw data and processed data. The difference between processed and unprocessed data can be observed from the Fig. 4a, b. The FpViz is an interactive tool used for large data (shown in Fig. 5) where nodes are represented with red color. Line between these nodes denotes connectivity. The connected path is the resultant frequent pattern obtained from the tool.

Textual form Graphical form Textual form Graphical form

A frequent pattern can also be visualized using Heatmaps [24]. Heatmaps are used to identify hot and cold regions. Figure 6 illustrates three different heatmaps: link position, click position, clicks/links. Herein, the links and clicks positions are visualized. Dimitrov et al. [24] divides the website screen into regions and gives

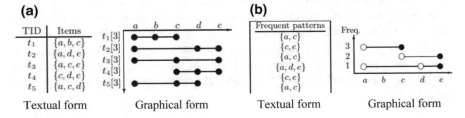

Fig. 4 a Representation of raw data navigations. **b** Representation of mining results of frequent patterns in both textual and graphical forms [23]

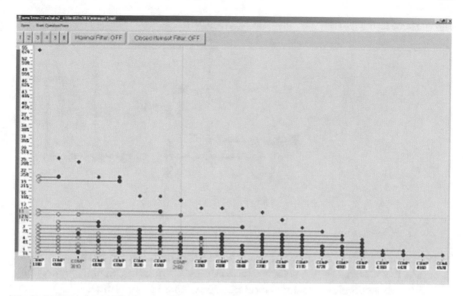

Fig. 5 FpViz module showing the visualization of frequent patterns mined [23]

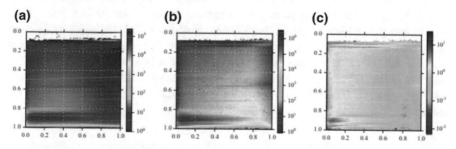

Fig. 6 Heatmaps [24]. **a** Links positions, **b** clicks positions, and (**c**) clicks/links

insights of frequently preferable regions by the users. They observed the region where links are placed. He analysed the region where users click on the links and found which regions is frequently (or rarely) clicked by the user. Figure 6a shows the position of links on the screen. This heatmap indicates high link density. Figure 6b presents heatmap of click position which indicates high click frequency regions. Figure 6c displays the number of clicks on a region. Here, dark (hot) colors indicate high frequency regions whereas light (cool) colors indicate a low frequency regions. Through heatmap, the author highlights that most of the users prefer to click on the left side of the screen.

(b) *Understanding Website Structure from Web Navigation Files*

Generally, a website structure is represented by the navigation tree. There are other layouts to visualize the navigation structure of the web or website. H-tree layouts (Fig. 7) representation are similar to binary trees which is suitable for balanced trees.

Fig. 7 H-tree layout [25]

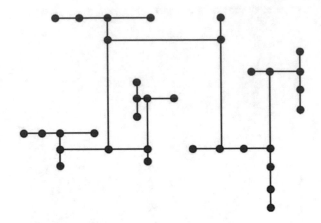

This representation is suitable for those websites which have a balanced tree-like structure. In the web, usually, websites have complex structure i.e., high connectivity. Therefore, this complex structure could not be obtained from H-tree.

To obtain a graphical representation of complex website structures Eades [26] and Chi et al. [27] suggested a variation of tree known as the radial tree or disk tree (Fig. 8a). The nodes are placed in the concentric circles based on tree depth. This representation is known as a radial tree. The node of the radial tree indicates the web page, and the edge indicates the hyperlink between two web pages. The root node is placed at the center, and concentric hierarchical circular layers with root are formed. Parent to children links is connected from inner to the outer circle. Using Disk tree, data layers can be obtained from the web log file which is helpful to generate user sessions.

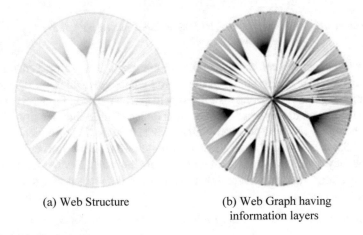

(a) Web Structure   (b) Web Graph having
information layers

Fig. 8 Visualizing Web Navigation Patterns [28]: **a** visualize the web structure. **b** Visualize super-position of web navigations on web structure

WebViz [28] is an interactive tool to analyse the web navigation data through the structure, pattern discovery on web navigation behavior. By default, in WebViz node size represents the frequency of page visit, node color shows average page view time, edge thickness indicates a number of the hyperlink, edge colour shows the percentage of hyperlink usage count (i.e., hyperlinks sharing the same start page divided by the total number of hyperlinks). Figure 8b displayed a web graph having several information layers. In addition to this, outputs of data mining models like classes or cluster of web pages or links can be visualized.

Figure 9 shows a closer look at the web navigation sequences using Radial Tree. While visualization two different sequences are observed. The first sequence shows a path 0->1->2, and the second sequence shows a path 0->3->4. These two sequences indicate the highest popular sequences navigated by the users. The frequent pattern traversed by the users can be determined through Radial Tree. The popularity or frequency of navigations is determined through support and confidence measures. These measures will be used to filter less frequent navigation pairs. The visualization of navigation pairs and frequently accessed navigation path is highlighted in the Fig. 9.

A tree representation of navigations can be viewed as the collection of sub-trees, to understand the hierarchal information structure cone tree [30] was defined. Cone tree projects the tree on a plane where sibling subtrees are attached to their parent node. Another simpler view of cone tree is known as the balloon tree. Figure 10 shows a balloon view of the navigations. Here, the root node is placed at the center. The group of sub-tree nodes is placed in the second layer. There are six sub-trees which are represented by the circles of nodes. However, a single child who is not a part of any subtree is represented as a node and directly linked to its parent node. Figure 10 indicates six child nodes. Using the balloon view of the cone tree, the position of nodes can be determined directly, without reference to the trees [31, 32].

(c) *Evolution of website structure with time*

With time new information is added to the website while outdated web pages are deleted. This updation of web site structure causes many changes in the website struc-

Fig. 9 Visualizing
sequences using radial tree
[29]

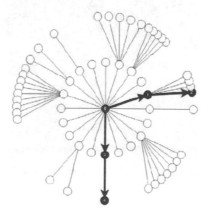

Fig. 10 Balloon view [30]

ture; i.e. adding and deleting the nodes or connections. Understanding the structural changes in the website becomes a major challenge for experts. To address this, Chi et al. [27] proposed a new visualization technique called the Time Tube. These tubes record several Disk tree over a period of time. This representation gives information about the change in website structure with time. Figure 11 illustrates four disk trees which show the evolution of web structure with time. This new visualization will guide the user to understand complex relationship and connection between production and consumption of information on the websites.

(d) *Identify the entry and exit point from the navigational structure*

The entry point is the web pages where users' typically enter the website. Exit points are the web pages where users' leave the website. These points provide insight into those web pages from which browsing starts and ends. To understand user entry and exit behavior, Chen et al. [33] had defined two operators: MINUS IN and MINUS OUT.

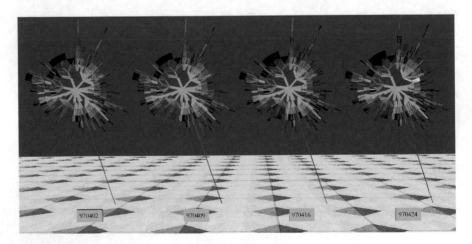

Fig. 11 Four disk trees shows the evolution of web struction with time [27]

To analysis, the entry point a MINUS IN operator is used which subtract the sum of the access to the page from any other pages that have a link to it from the number of visits of that page. This difference represents how often users enter the page directly. To analysis the exit point a MINUS OUT operator is used which subtract the sum of the access to any other pages from the specific page from the number of visits of that page. This difference indicates how often a user stops the navigation and leaves the site. Figure 12 illustrates the process to get Entry Points and Exit Points of the website.

(e) *Visualize the traffic and popularity of the Website(s) from Web Server logs*

The traffic on a web grows exponentially with time. A website having more traffic indicates its popularity among other websites. Understanding the traffic will give the insight of the incoming and outgoing users on the website. In addition, it provides information about the popularity of the website. Mark [34] presented some plots to understand the distribution of web traffic i.e., in-degree and out-degree of the node. According to him, the in-degree and out-degree must satisfy the power law (shown in Fig. 13a, b). This distribution depicts the strength of incoming and outgoing degree. Furthermore, the distribution is used to understand user traffic visiting the website or leaving the website. Figure 14a presents inbound traffic of the web server users. Figure 14b presents outbound traffic of the users contributing to the traffic of the web server using power law. These distributions measure popularity of a Website and stated that most popular sites have unbounded audience.

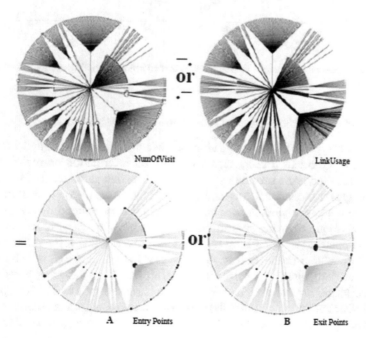

Fig. 12 Process to obtain entry and exit points using disk tree [33]

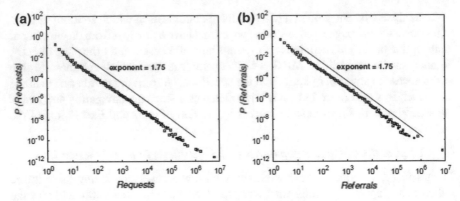

Fig. 13 Distributions of **a** in-degree strength and **b** out degree strength the web server data [34]

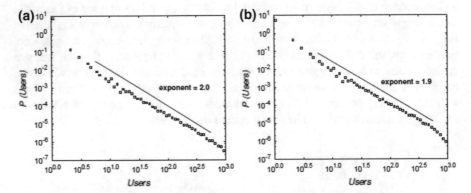

Fig. 14 Distributions of **a** unique incoming users and **b**) unique outgoing users from web server data [34]

(f) *Classifying searching-oriented user or browsing-oriented user*

There is two types of users navigation behavior: search-oriented and browsing-oriented. Search-oriented users search the desired information on the web while browsing-oriented users surf the web randomly. The navigation data of users gives useful insights to classify these users. Figure 15a shows the distribution of the active user's requests per second. A user behaviour can be understood by a ratio of the number of unique referring sites to the unique target sites. User browsing behavior can be obtained by comparing referring host and servers. If the referring hosts is less as compared to the servers, then the user browses through search engines, social networking sites, or a personal bookmark file. If the referring hosts is high compared to the servers, then a user is having surfing behavior: Fig. 15b presents bimodal distribution which states the existence of two user groups: search oriented and browsing oriented. Search-oriented users visit more sites as compared to surfers for each referrer.

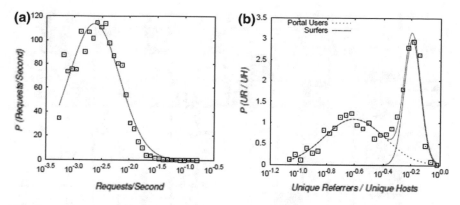

Fig. 15 a Distribution of the requests made per second by each user using log-normal fit. **b** Distribution of the ratio of the unique referring sites to unique target sites for each user using bimodal distribution [34]

Catledge and Pitkow [35] categorized search oriented browsing and surfer browsing into three using user browsing behavior. These three dominant types of behavior are still applied in many research areas. The first browsing behavior is known as search browsing, which is goal specific browsing. This is a directed search with the specified goal. The second browsing behavior is referred to as general purpose browsing. The idea of the browsing goal exists, but the path to the goal is undefined. In general purpose browsing, the user consults different sources, which are supposed to contain the desired information. The third browsing type is serendipitous browsing which is truly random browsing behavior. Figure 16 illustrates the type of users based on their browsing behavior [36]. The search browser is known as goal-oriented users, and the serendipitous browser is known as exploratory users. Figure 17 depicts an example of the goal-orient browser. Herein, the user follows a path towards blogs. There is less variation in the navigation path. Hence, the user is categorized into the goal-oriented browser. Figure 18 illustrates the exploratory browsing behavior. In this typical user browse a mixture of interesting content i.e., stories or blogs and ends in the comments section frequently.

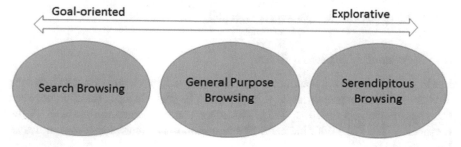

Fig. 16 Type of users in online browsing [36]

		1	2	3	4	5	6	7	8	9	10	11	12	13
Editorial content	Front page [N]	█					█	█	█		█		█	
	Stories [E]				█									
Community content	Blogs [B]		█			█	█	█		█		█		
	Video Blogs [V]													
	Social Media [M]													
	Comments [C]													
	Alternative Blogs [T]			█										
Commercial	Ads [A]													
	Stores [S]													

Fig. 17 Navigation path for goal-oriented browser [36]

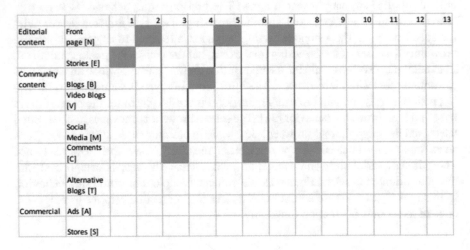

Fig. 18 Navigation path for exploratory browser [36]

(g) *Correlation of Sessions or Users*

Finding a correlation between users or sessions or features is important to find how similar or dissimilar properties they have. Figure 19 presents the correlation between the features using a heat map [37]. The color code indicates the strength of the correlation. The red colour indicates high correlation whereas blue colour indicates low correlation.

(h) *Noise Detection*

Fig. 19 Correlations of the features [37]

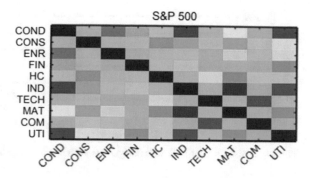

Data visualization gives a view to identify the normal and unusual users easily. Mark et al. [34] examines click stream sizes of the users'. Distribution of empty-referrer request indicates the user frequency who has jumped directly to a specific page (e.g., a homepage, a bookmark, etc.) instead of following web pages from the visited web pages. The resulting distributions are shown in Fig. 20a, b.

In web navigation, the user can either follow a forward path or backward path. The backward path shows the presence of noise as navigation doesn't follow the website topology. Forward path occurs when the user is interested in desired information from the web pages which has not been traversed. However, the backward path occurs when the user attempts to follow visited web pages. Jindal et al. [20] presented a study on backward navigations over varied session length. They have examined how frequently the user follows backward navigation path. According to them, the backward path are the repeated sequence of web pages in the session. The presence of a backward path produce longer sessions and has noise. For instance, a navigation path E->B->J-> H->B has noise as it consists backward navigation from H to B and it does not follow web topology (there is no link from H to B in the website structure).

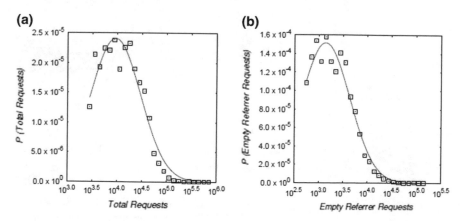

Fig. 20 Distributions of the **a** number of requests per user **b** number of empty-referrer requests per user [34]

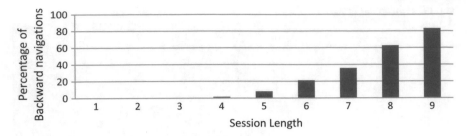

Fig. 21 Distribution of backward navigations with session length [20]

Figure 21 shows the number of backward navigations increases with session length [20]. The sessions with length one and two have no backward navigations. The backward navigations of higher order consist more repetition ranging from 0 to 3 which means session in the dataset may consist 0, 1, 2 or 3 cycles. This analysis gives insight to the users about variations of backward navigations with session length. This implies that user with more session length usually performs backward navigations. Presence of the backward paths makes prediction models difficult to learn user intentions.

An anomaly detection tool, OPAvion [38] is composed of three sub-parts: Data Summarization, Anomaly Detection, and Visualization. Summarization gives an overview of large network structures and their features. Anomaly detection phase extract features of the graph, node and neighborhood information. The graph induced with this information forms egonet. To search for the anomaly, the majority of normal neighboring nodes should be understood. The deviation (if any) has been recorded and analyzed to capture anomaly. Figure 22a the 'Stack Overflow' Question & Answer network is illustrated as Egonet. Redline is the least squares on the median values (dark blue circles) of each bin. The top anomalies deviating are marked with trian-

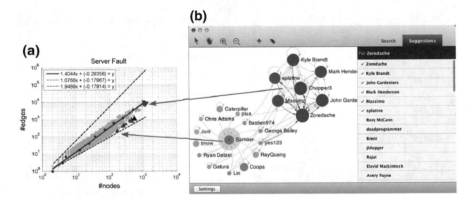

Fig. 22 a Illustrating Egonet of the 'Stack Overflow' Q&A graph. The top anomalies deviating are marked with triangles. **b** Visualization module working with the anomaly detection module, showing as a star [38]

gles. Figure 22b presents the anomaly detection visualization module. The anomaly (Sameer, at its center) is presented with the star in Fig. 22b. The Nodes represent Stack Overflow users and a directed edge from a user who asks a question, to another user who answers it. The user Sameer is a maven who has answered a lot of questions (high in-degree) from other users, but he has never interacted with the other mavens in the near-clique who have both asked and answered numerous questions. Therefore, Sameer is the anomaly of this network.

2.4 Pattern Discovery

Pattern discovery has attracted many experts to find interesting patterns, to understand user behaviour, etc. Wide research has been carried out to find interesting patterns varied according to the application. Hu et al. [39], Pani et al. [40], Singh and Singh [41], Facca and Lanzi [42], Jindal and Sardana [43] present an overview of web usage mining system and survey on pattern discovery techniques. These papers present the detail of the components of web navigation mining. Each component requires a considerable attempt to accomplished desired objectives. The pattern discovery component is the most important part of the web navigation system. Several data mining techniques were proposed like clustering, classification, association rule mining, sequential mining used for Web Navigation Mining. Tseng et al. [44] proposed association rule-based techniques to discover temporal patterns using temporal properties like weekdays or hours. Chordia and Adhiya [45] proposed the sequence alignment method for clustering sessions based on certain criteria. Abrishami et al. [46] and The [47] integrate three web navigation prediction techniques and formed new hybrid approaches. The proposed techniques combine association rule mining, clustering and sequence mining techniques to improve the accuracy of the model. It has been seen that the Markov Model is extensively used to discover web navigation patterns. Markov models [48–54] are stochastic processes and well suited for modelling and analyzing sequential patterns. The detail description of Markov based models is highlighted by Sunil et al. [55] and Jindal [56]. Based on application and objectives Markov models are combined with data mining techniques. Thwe [47] and Vishwakarma et al. [57] analyzed the application of Markov based Models on clusters. Sampath et al. [58–60] present the integrated model to improve the prediction accuracy. They observed the effect of association rule mining on Markov based Models.

2.5 Post Pattern Discovery Visualization: Stage 2

The second stage of visualization is used to visualize the interesting patterns obtained from pattern discovery phase. It help to understand user navigation behavior after data mining algorithms.

(a)   **(b)**

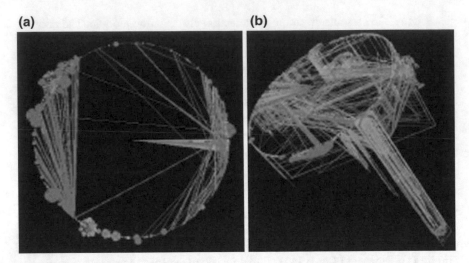

Fig. 23 **a** 2D visualization of web logs using Strahler coloring technique. **b** 3D visualization of more information which clarifies user behavior in and between clusters [61]

(a) *Visualizing the user behaviour after clustering*

To visualize the users and their connections in and out to the clusters, Amir et al. [61] presents the visualization of weblog using Strahler numbering (Fig. 23a).

Strahler number provides quantitative information of the tree shape and complexity by assigning colours to the tree edges. The leaf node has Strahler number one. The node having one or more childs have Strahler number K or K + 1 depending upon the strahler number of their childrens. Herman et al. [25] described how Stracher numbers can be used to provide information about tree structure. The cylinder part of Fig. 23b gives insights of web usage visualization of users in and between the cluster. The Center node of the circular basement is the first page of the web site from which users scatter to different clusters of web pages. The Red spectrum denotes the entry point into clusters and Blue spectrum illustrates the exit point of the users.

Figure 24 indicates that the combination of some visualizations techniques can be used to find new patterns and analyze the patterns obtained through mining algorithms. Through this framework, frequent patterns can be extracted and visually superimposed as a graph of node tubes. The thickness of these tubes indicates aggregated support of navigated sequences. The size of clickable Cube Glyphs [61] indicates the hit of web pages. In Fig. 24b shows the superimposition of Weblogs on top of Web Structure with higher order layout. The top node represents the first page of the website. This hierarchical output of layouts makes analysis easier.

(d) *Understanding Collaborative Customers Buying Patterns*

The rules discovered by pattern discovery phase can be visualized to understand customers buying patterns [62, 63]. Figure 25 presents MineSet's Association Rule Visualizer [62] which maps features of the market basket in x-axis and y-axis. The

(a) **(b)**

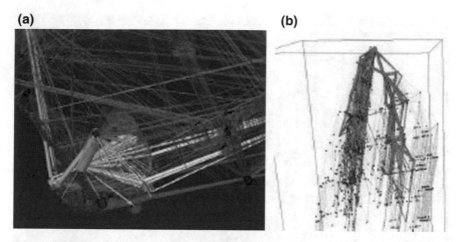

Fig. 24 **a** Frequent access patterns extracted by web mining process using *confidence threshold* are visually superimposed as a graph of node-tubes (as opposed to node-links). **b** Superimposition of web usage on top of web structure with higher order layout [61]

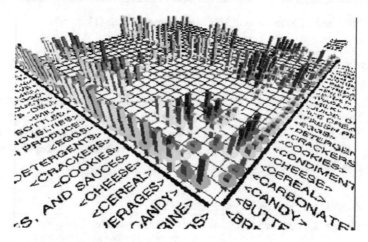

Fig. 25 MineSet's association rule visualizer [62] maps rules to the x- and y-axes. The confidence is shown as the bars and the support as the bars with discs

confidence of association rule is highlighted with bar whereas support as the bar with the disc. The colour on the bar signifies the importance of the rule.

Sequences Viewer [38] helps experts to navigate and analyze the sequences identified by pattern discovery techniques (see Fig. 26). The Point Cloud representation is a sequence viewer which allows users to visualize groups of sequences. Through this view, the centers of the groups, the distance from the centers, and associated sequences can be easily determined.

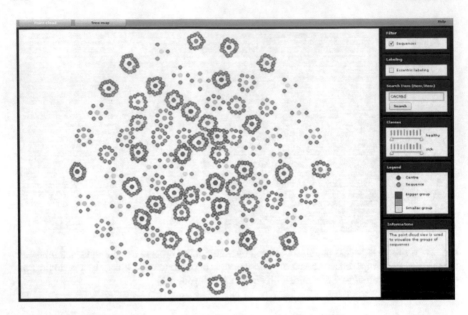

Fig. 26 Point cloud with sequences and highlighted searched items [38]

3 Research Questions

This section highlights research challenges which mainly occurred in web navigation mining. The research challenges presented visualization aspect in web usage mining. Table 1 categorizes visualization aspects into two: (a) Web Usage Mining and (b) Web Structure Mining through navigation files. The solution to these challenges is described in Sect. 2.

Table 1 Visualization aspects in web navigation mining

Visualization aspects from web navigation mining	Visualization aspects from web structural mining through navigation files
(a) How has information been accessed?	(a) How many web pages has been added? Modified? Deleted?
(b) How frequently?	
(c) How recently?	(b) How is new web page accessed? When does it become popular? When does it die?
(d) What is the most and least popular information?	
(e) Which web page user follows to enter the site? exit?	(c) How does the addition of web page change the navigation patterns within the site? Can people still navigate to desired web page?
(f) In which region user spend time?	
(g) How long do they spend there?	
(h) How do people travel within site?	(d) Do people look for deleted web page? How relevant is the deleted web page?
(i) What is the most and least popular path?	
(j) Who are the users visiting the website?	

4 Visualization Tools

The tools used for Visualization has been described in this section. There is a wide range of data mining tools available which presents different visualization aspects. Following Tables 2, 3 and 4 presented visualization tools in data pre-processing, pattern discovery and pattern analysis.

Table 2 Data Pre-processing tools

Tools	Features
Data preparator [64]	It is free software used for the common task in pattern mining. It performs data filtering, pattern discovery and data transformation
Sumatra TT [65]	It is a platform independent data transformation tool
Lisp miner [66]	Analyzing the navigations and preforms pre-processing
Speed tracer [67]	It mines web log files, visualizes and identify the sessions to reconstruct user navigational sessions

Table 3 Pattern discovery tools

Tools	Features
Sewebar- Cms [68]	Selected rules among large set rules discovered from association rule mining. It provides interaction between data analyst and domain expert to discover patterns
i-Miner [69]	Discover cluster through the fuzzy clustering algorithm which is further used for pattern discovery and analysis
Argunaut [70]	Develop interesting patterns by using the sequence of various rules
MiDas (Mining Internet Data for Associative Sequences) [71]	It is an extension of traditional sequential techniques which adds more features to the existing techniques. It discovers market-based navigation patterns from log files

Table 4 Pattern analysis tools

Tools	Features
Webalizer [72]	Discover web pages after analyzing the patterns
Naviz [73]	Visualization tool which combines the 2-D graph of user navigation patterns. It clusters the related pages and describes the pattern of user navigation on the web
WebViz [74]	Analyze the patterns and provides a graphical form of the patterns
Web miner [75]	Analyse useful patterns and provides the user specific information
Stratdyn [76]	Enhances WUM and provides patterns visualization

5 Conclusion

Visualization can be performed to analyze Web Navigations with a different perspective. The chapter presents the framework for having various components of web navigation mining. The components are weblogs consisting of users navigation details, Pre-processing and Cleaning and Pattern Recognition. In addition, two components are depicting the visualization of web data after preprocessing and after identification of patterns. The visualization can be performed to investigate the website structure, evolution of the website, user navigation behavior, frequent and rare patterns, outlier or anomaly detection, etc. It also determines and analyses the interesting patterns from the clusters or rules. It validates the quality of patterns discovered. Since the web is the largest repository of information which makes users difficult to find the desired information. Visualization is used to provide a summarised view of large data pictorially. Visualization can be used to analyze and discover hidden information from large data. The application of visualization and its metrics varies according to the requirements. The chapter discusses various visualization aspects in the area of web navigation mining. In the end, some tools are listed which makes analysis simpler and easier.

References

1. Flavián C et al (2016) Web design: a key factor for the website success. J Syst Inf Technol
2. Rahi P et al (2012) Business intelligence: a rapidly growing option through web mining. IOSR J Comput Eng (IOSRJCE) 6(1):22–29
3. Jindal H et al (2018) PowKMeans: a hybrid approach for gray sheep users detection and their recommendations. Int J Inf Technol Web Eng 13(2):56–69
4. Kazienko P et al (2004) Personalized web advertising method. In: International conference on adaptive hypermedia and adaptive web-based systems, pp 146–155
5. Ketukumar B, Patel AR, Web data mining in e-commerce study, analysis, issues and improving business decision making. PhD Thesis, Hemchandracharya North Gujarat University, Patan, India
6. McKinney W (2018) Python for data analysis data wrangling with pandas, NumPy, and IPython. O'Reilly
7. https://www.loganalyzer.net/log-analysis-tutorial/what-is-log-file.html, Accessed 13 Feb 2019
8. https://www.ibm.com/support/knowledgecenter/en/ssw_ibm_i_71/rzaie/rzaielogformat.htm, Accessed 14 Feb 2019
9. https://en.wikipedia.org/wiki/Common_Log_Format, Accessed on 14 Feb 2019
10. http://www.herongyang.com/Windows/Web-Log-File-IIS-Apache-Sample.html, Accessed 13 Feb 2019
11. Masand B, Spiliopoulou M (eds) Advances in web usage analysis and user profiling. In: LNAI 1836. Springer, Berlin, Germany, pp 163–182
12. Catledge L, Pitkow J (1995) Characterizing browsing behaviors on the world wide web. Comput Netw ISDN Syst 26:1065–1073
13. Cooley R, Mobasher B, Srivastava J (1999) Data preparation for mining world wide web browsing patterns. J. of Knowl Inf Syst 1, 5–32
14. Cooley R, Tan P, Srivastava J (2000) Discovery of interesting usage patterns from web data

15. Myra S, Bamshad M, Bettina B, Miki N (2003) A framework for the evaluation of session reconstruction heuristics in web-usage analysis. Inf J Comput 15(2):171–190

16. Murat AB, Ismail HT, Murat D, Ahmet C (2012) Discovering better navigation sequences for the session construction problem. Data Knowl Eng 73:58–72

17. Dell RF, Roman PE, Valaquez JD (2008) Web user session reconstruction using integer programming. In: IEEE/WIC/ACM international conference on web intelligence and intelligence agent technology, vol 1, pp 385–388

18. Dell RF, Roman PE, Valaquez JD (2009) Fast combinatorial algorithm for web user session reconstruction. IFIP

19. Dell RF, Roman PE, Valaquez JD (2009) Web user session reconstruction with back button browsing. In: Knowledge-based and intelligence information and engineering systems, of Lecture notes in Computer Science, vol 5711. Springer, Berlin, pp 326–332

20. Jindal H, Sardana N (2018) Elimination of backward browsing using decomposition and compression for efficient navigation prediction. Int J Web Based Commun (IJWBC) 14(2)

21. Zhang J, Ghorbani AA (2004) The reconstruction of user sessions from a server log using improved time-oriented heuristics. In: Annual conference on communication networks and services research, pp 315–322

22. Chen Z et al, Linear time algorithms for finding maximal forward references. In: Information technology: coding and computing [Computers and Communications] proceedings. ITCC, pp 160–164

23. Leung CK-S, FpVAT: a visual analytic tool for supporting frequent pattern mining. SIGKDD Explor 11(2)

24. Dimitrov D et al (2016) Visual positions of links and clicks on wikipedia. WWW, ACM

25. Herman I, Melançon G (2000) Graph visualization and navigation in information visualization: a survey. IEEE Trans Vis Comput Graph 6

26. Eades P (1992) Drawing free trees. Bull Inst Comb Appl 10–36

27. Chi EH, Pitkow J, Mackinlay J, Pirolli P, Gossweiler R, Card SK (1998) Visualizing the evolution of web ecologies. In: Proceeding of CHI

28. Chen J, Zheng T, Thorne W, Zaiane OR, Goebel R, Visual data mining of web navigational data

29. Oosthuizen C et al (2006) Visual web mining of organizational web sites. In: Proceedings of the information visualization

30. Carrière J, Kazman R (1995) Research report: interacting with huge hierarchies: beyond cone trees. In: Proceedings of the IEEE conference on information visualization '95. IEEE CS Press, pp 74–81

31. Melançon G, Herman I (1998) Circular drawings of rooted trees. Reports of the Centre for Mathematics and Computer Sciences, Report number INS–9817. http://www.cwi.nl/InfoVisu/papers/circular.pdf

32. Yuntao J (2007) Drawing trees: how many circles to use?

33. Chen J et. al (2004) Visualizing and discovering web navigational patterns. In: Seventh international workshop on the web and databases

34. Mark M, What's in a session: tracking individual behavior on the web. ACM

35. Catledge L, Pitkow J (1995) Characterizing browsing strategies in the world-wide web. Comput Netw ISDN Syst 27:1065–1073

36. Lindén M (2016) Path analysis of online users using clickstream data: case online magazine website

37. Kenett DY et al (2014) Partial correlation analysis. applications for financial markets. Quant Financ

38. Sallaberry A et. al (2011) Sequential patterns mining and gene sequence visualization to discover novelty from microarray data. J Biomed Inform 44, 760–774

39. Hu C et al (2003) World wide web usage mining systems and technologies. J Syst Cybern Inform 1(4):53–59

40. Pani SK et al (2011) Web usage mining: a survey on pattern extraction from web logs. Int J Instrum Control Autom 1(1):15–23

41. Singh B, Singh HK (2010) Web data mining research: a survey. In: IEEE international conference on computational intelligence and computing research (ICCIC), pp 1–10
42. Facca FM, Lanzi PL (2005) Mining interesting knowledge from weblogs: a survey. Data Knowl Eng 53(3):225–241
43. Jindal H, Sardana N (2007) Empirical analysis of web navigation prediction techniques. J Cases Inf Technol (IGI Global) 19(1)
44. Tseng VS, Lin KW, Chang J-C (2008) Prediction of user navigation patterns by mining the temporal web usage evolution. Soft Comput 12(2), 157–163
45. Chordia BS, Adhiya KP (2011) Grouping web access sequences using sequence alignment method. Indian J Comput Sci Eng (IJCSE) 2(3):308–314
46. Abrishami S, Naghibzadeh M, Jalali M (2012) Web page recommendation based on semantic web usage mining. Social informatics. Springer, Berlin, pp 393–405
47. Thwe P (2014) Web page access prediction based on integrated approach. Int J Comput Sci Bus Inform 12(1):55–64
48. Papoulis A (1991) Probability, random variables, and stochastic processes. McGraw-Hill, USA
49. Bhawna N, Suresh J (2010) Generating a new model for predicting the next accessed web pages in web usage mining. In: IEEE international conference on emerging trends in engineering and technology, pp 485–490
50. Xu L, Zhang W, Chen L (2010) Modelling users visiting behaviours for web load testing by continuous time Markov chain. In: IEEE 7th web information systems and application conference. IEEE, pp 59–64
51. Wang C-T et al (2015) A stack-based Markov model in web page navigability measure. In: International conference on machine learning and cybernetics (ICMLC), vol 5. IEEE, pp 1748–1753
52. Speiser M, Antonini G, Labbi A (2011) Ranking web-based partial orders by significance using a Markov reference mode. In: IEEE international conference on data mining, pp 665–674
53. Dhyani D, Bhowmick SS, Ng W-K (2003) Modelling and predicting a web page accesses using Markov processes. In: Proceedings 14th international workshop on database and expert systems applications. IEEE
54. Dongshan A, Junyi S (2002) A new Markov model for web access prediction. IEEE Comput Sci Eng 4(6):34–39
55. Sunil K, Gupta S, Gupta A (2014) A survey on Markov model. MIT Int J Comput Sci Inf Technol 4(1):29–33
56. Jindal H, Sardana N (2018) Decomposition and compression of backward browsing for efficient session regeneration and navigation prediction. Int J Web Based Commun (Inderscience) 14(2)
57. Vishwakarma S, Lade S, Suman M, Patel D (2013) Web user prediction by: integrating Markov model with different features. Int J Eng Res Sci Technol 2(4):74–83
58. Sampath P, Ramya D (2013) Analysis of web page prediction by Markov model and modified Markov model with association rule mining. Int J Comput Sci Technol
59. Sampath P, Ramya D (2013) Performance analysis of web page prediction with Markov model, association rule mining (ARM) and association rule mining with statistical features (Arm-Sf). IOSR J Comput Eng 8(5):70–74
60. Sampath P, Wahi A, Ramya D (2014) A comparative analysis of Markov model with clustering and association rule mining for better web page prediction. J Theor Appl Inf Technol 63(3):579–582
61. Amir H et. al (2004) Visual web mining. WWW. ACM
62. S. G. Inc. Mineset (2001) Mineset. http://www.sgi.com/software/mineset
63. Hofmann H, Siebes A, Wilhelm A (2000) Visualizing association rules with interactive mosaic plots. In: SIGKDD International conference on knowledge discovery & data mining (KDD 2000), Boston, MA
64. http://www.datapreparator.com/
65. Stěpankova O, Klema J, Miksovsky P (2003) Collaborative data mining with Ramsys and Sumatra TT, prediction of resources for a health farm. In: Mladenić D et al (ed) Data mining and decision support
66. https://lispminer.vse.cz/, Accessed 9 Mar 2019

67. https://code.google.com/archive/p/speedtracer/, Accessed 8 Mar 2019
68. Kliegr T, SEWEBAR-CMS: semantic analytical report authoring for data mining results
69. Abraham A (2003) i-Miner: a web usage mining framework using hierarchical intelligent systems. In: The 12th IEEE international conference on fuzzy systems
70. https://www.researchgate.net/figure/ARGUNAUTs-Moderators-Interface-and-some-of-its-shallow-alerts_fig12_220049800
71. Büchner AG, Navigation pattern discovery from internet data
72. http://www.webalizer.org/
73. https://navizanalytics.com/, Accessed 9 Mar 2019
74. Pitkow JE, Bharat KA, WEBVIZ: a tool for world-wide web access log analysis. In: Proceedings of first international WWW Conference
75. https://www.crypto-webminer.com/
76. Berendt B et al (2001) Visualizing individual differences in Web navigation: STRATDYN, a tool for analyzing navigation patterns. Behav Res Methods Instrum Comput 33(2):243–57

Research Trends for Named Entity Recognition in Hindi Language

Arti Jain, Devendra K. Tayal, Divakar Yadav and Anuja Arora

Abstract Named Entity Recognition (NER) is a process of identification and classification of names into pre-defined categories- person, location, organization, date, time and others. NER serves as one of the most valuable application tools for varying languages and domains. Despite its popularity and successful deployment in English, this area is still exploratory for the Hindi language. NER in Hindi is a challenging task due to the scarceness of language resources and complex morphological structure. An initial impetus to write this chapter constitutes concise research on prevailing NER systems in the Hindi language. To achieve this objective, all-inclusive review and analysis of the research trends for NER in Hindi starting from the year 1999 to till date is conducted from varied articles. These articles include full-length articles, survey papers, dissertations, and guidelines which exploit decision analysis of the NER with respect to six vital aspects. All these aspects are collaborated and visualized in details within the chapter. The future research directions, challenges, and open issues in the field of NER are also presented for keen researchers. One can then design a unified, coherent NER system which can be applied to vivid language processing tasks for the Hindi and other Indian languages.

Keywords Named entity recognition · Hindi language · Training corpora · Ner techniques · Gazetteer lists · Evaluation measures

A. Jain (✉) · A. Arora
CSE, Jaypee Institute of Information Technology, Noida, UP, India
e-mail: arti.jain@jiit.ac.in

A. Arora
e-mail: anuja.arora@jiit.ac.in

D. K. Tayal
CSE, Indira Gandhi Delhi Technical University for Women, Delhi, India
e-mail: dev_tayal2001@yahoo.com

D. Yadav
CSE, Madan Mohan Malvia University of Technology, Gorakhpur, UP, India
e-mail: dsycs@mmmut.ac.in

© Springer Nature Switzerland AG 2020
J. Hemanth et al. (eds.), *Data Visualization and Knowledge Engineering*,
Lecture Notes on Data Engineering and Communications Technologies 32,
https://doi.org/10.1007/978-3-030-25797-2_10

1   Introduction

Named Entity Recognition (NER) [2, 31, 39, 44, 56, 57, 68, 79–81] is a sequence labelling task which seeks identification and classification of Named Entities (NEs). An identification of named entity represents the presence of a name- word or term or phrase as an entity within a given text. Classification of named entity denotes the role of an identified NE such as person name, location name, organization name, date, time, distance, and percent. In general, named entities are distributed into three universally [5] defined tags (Fig. 1)—ENAMEX, TIMEX, and NUMEX which are detailed below:

- **ENAMEX Tag**: ENAMEX tag is used for names entities e.g. person, location, organization, facility, locomotive, artifact, entertainment, material;
- **TIMEX Tag**: TIMEX tag is used for temporal entities e.g. time, day, month, period, date, special day, year;
- **NUMEX Tag**: NUMEX tag is used for numerical entities e.g. distance, quantity, money, count, percent, measurement.

   In addition, depending upon the specific user needs there are marginal NE types viz. book title, phone number and email address [91], medical emails, scientific books and religious text [54], job title [8], scientist and film [27], project name and research area [92], device, car, cell phone, currency [64].

   NER is recommended in situations where NEs have more importance than the actions they perform. Over the decades, NER is proved successful in vivid natural language processing application areas such as information retrieval [47], text summarization [35], co-reference resolution [11], word sense disambiguation [55], question answering [63], machine translation [90] and so on. NER has a profound impact on the society as it enables crowdsourcing [4] which is rapidly growing as social media

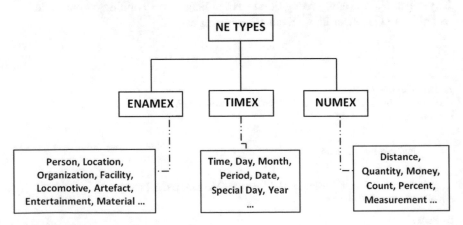

Fig. 1 NE categories and their sub-categories

content through plug-in of GATE[1] crowdsourcing and DBpedia[2] linking. In actuality, NER is a first step towards extraction of structured information from an unstructured text- newspaper articles, web pages and many more.

For more than a decade, NER for Indian languages, especially Hindi is thriving as a promising research topic. Several researchers have contributed their work for NER in Hindi using statistical [82, 88], rule-based [76] and hybrid systems [36, 77]. Some well-known statistical Hindi NER techniques are- Conditional Random Field (CRF) [52, 82], Maximum Entropy (MaxEnt) [67], Hidden Markov Model (HMM) [7], Expectation Maximization (EM) [9], Support Vector Machine (SVM) [20, 73], Genetic Algorithm (GA) [21, 22], Long Short Term Memory (LSTM) [1], Hyperspace Analogue to Language (HAL) [41, 42]. Some rule-based Hindi NER techniques are Association Rule Mining (ARM) [40], rule list-lookup [46], memory-based learning [76]. Some hybrid Hindi NER techniques are- combination of MaxEnt, rules and gazetteers [66], dimensionality reduction- CRF and SVM [74], Multi-Objective Optimization (MOO) [23, 75], K-Nearest Neighbour (KNN) and HMM [77], CRF and LSTM [87], HAL and CRF [41, 42]. Such a situation has probably contributed to the fact that the Hindi language has a wider perspective, a wider range of content, and so deserves more attention towards language-based research. The Hindi language is in the Devanagari [34] script and comes at the fourth position as the world's most-spoken first language. Also, it is characterized by the highly inflectional, morphologically rich, suffix-based, and word re-ordering language. A commendable source for the Hindi language processing is the Hindi WordNet[3] (HWN).

This chapter presents a comprehensive understanding of NER in the Hindi language and related issues. To achieve this objective, all-inclusive review and analysis of the research trends for NER in the Hindi starting from the year 1999 to till date is conducted from varied articles. These articles include full-length articles, survey papers, dissertations, and guidelines while exploiting NER decision analysis w.r.to six vital aspects. These research aspects are, namely- datasets, state-of-art techniques, training corpora, NE types, gazetteer lists, and evaluation measures. All these Hindi NER aspects collaborate among each other that envisage researchers to answer the following research questions:

- What are the existing datasets that are available for the Hindi NER?
- What are the different techniques that are applied for the Hindi NER?
- What are the varied training corpora that are accessible for the Hindi NER?
- What are the current and emerging named entity types for the Hindi NER?
- What are the several gazetteers lists that are available for the Hindi NER?
- What are the performance evaluation metrics to evaluate Hindi NER?
- What are the challenges, future directions and open issues for the Hindi NER?

---

[1] https://gate.ac.uk/wiki/crowdsourcing.html.

[2] https://github.com/dbpedia/links.

[3] http://www.cfilt.iitb.ac.in/wordnet/webhwn/.

In order to respond to the above questions, relevant research articles on the named entity recognition in the Hindi language are to be mined while laying down the following criteria: -

- **Level of relevance**: The research articles are given the highest to the lowest level of relevance based upon the SCImago Journal Ranking, author level metric/H-index, impact factor, citation and so on. For example, an article on the Hindi NER which is published in an International Journal with high impact factor and H-index has a higher relevance.
- **String-based search**: The string-based search is performed mainly on the titles of the research articles. For example, <NER for South Asian Languages>, <NER for South and South Asian Languages>, <NER for Indian Languages>, <Indian Language NER>, <NER in Indian Languages>, <NE in Under-Resourced Language>, <Hindi Language NER>, <Hindi NER>, <NER in Hindi>, <NER for Hindi>, <Hindi NE Annotation>.
- **Research timeline**: The timeline of the online Hindi NER research articles that starts from a novice study [9] to the latest [62] are been taken care of. NER for the Indian languages have come to limelight during 2007–2008 and since then research in this field is marching ahead.

Rest of the chapter is organized as follows. Section 2 describes named entity tagset for the Hindi NER. Section 3 describes state-of-art for NER in Hindi. Section 4 illustrates the contemporary challenges in Hindi NER. Section 5 concludes the chapter.

2   Named Entity Tagset for the Hindi Language

Named entity tagset for the Hindi language is standardized using the tagset guidelines[4] that are prepared at the AU-KBC Research Centre, Chennai, India. According to these guidelines, there are three prime NE categories for the Hindi NER viz. ENAMEX, NUMEX, and TIMEX which have a certain resemblance to the English NER guidelines.[5] Further, each of these NE categories contains various NE types that are described here using XML@ script.[6]

2.1   ENAMEX Tag

ENAMEX entity for the Hindi language includes various name expressions such as person, location, organization, facility, locomotive, artefact, entertainment, cuisine, organism, plant, disease NEs. All such NEs are listed here one by one.

---

[4]www.au-kbc.org/.

[5]https://cs.nyu.edu/faculty/grishman/mu6.html.

[6]https://www.w3schools.com/xml/.

2.1.1 Person NE

Person entity is considered as a human being, who is an individual or a set of individuals/group. It also includes fictional characters from the story, novel, etc. A person name includes person first name, middle name, and last name, along with titles such as Mr., Mrs., Ms., and Dr.

E.g. <ENAMEX TYPE = "PERSON"> जशोदाबेन नरेंद्र मोदी</ENAMEX>

English Translation: जशोदाबेन नरेंद्र मोदी(Jashodaben Narendra Modi)

2.1.2 Location NE

Location entity is considered as a geographical entity such as geographical area, landmass, water body and geological formation. It includes address such as PO Box, street name, house/plot number, area name, and pin code. It also includes name of village, district, town, metropolitan city, cosmopolitan city, state or province capital, state, the national capital, nation, highway, continent, any other place such as religious place, etc. In addition, it includes the nickname given to a city or state such as Jaipur-Pink City. Further, fine-grained refinement of few locations based NEs also persist such as water body includes the name of pond, lake, stream, river, sea, and ocean. The landscape includes mountain, mountain range, valley, glacier, desert, forest, ghat, and wetland and so on.

E.g. <ENAMEX TYPE = "LOCATION"> पुरानी दिल्ली</ENAMEX>

English Translation: पुरानी दिल्ली(Old Delhi)

2.1.3 Organization NE

Organization entity is considered as limited to a corporation, agency, and another group of people that are defined by an established organizational structure. It includes the name of the village panchayat, municipality, road corporations, travel operator, advertisement agency, and TV channels such as news channel; political party/group, the militant/terrorist organization, professional regulatory bodies such as Indian Medical Association, IEEE; charitable, religious board and many more.

E.g. <ENAMEX TYPE = "ORGANIZATION"> माता वैष्णो देवी तीर्थ बोर्ड</ENAMEX>

English Translation: माता वैष्णो देवी तीर्थ बोर्ड(Mata Vaishno Devi Shrine Board)

2.1.4 Facility NE

Facility entity is considered as limited to building, man-made structure, and real-estate such as hospital, institute, library, hotel, factory, airport, railway station, police station, fire station, harbor, public comfort station, etc. The hospital can include the name of a hospital, clinic, dispensary, primary health unit. Institute can include the name of an educational institution, research center, and training center. The library can include the name of a public library, private library. The hotel can include the name of a hotel, restaurant, lodge, fast food center. Factory can include the name of a factory, refinery, chemical plant, sewage treatment plant. The airport can include the name of the domestic airport, international airport, air force base station and so on.

E.g. <ENAMEX TYPE = "FACILITY"> इन्दिरा गाँधी अन्तरराष्ट्रीय हवाईअड्डा </ENAMEX>

English Translation: इन्दिरा गाँधी अन्तरराष्ट्रीय हवाईअड्डा(Indira Gandhi International Airport)

2.1.5   Locomotive NE

The locomotive entity is considered as a physical device that is designed to move, carry, pull or push an object from one location to another. It includes the name of flight, train, bus and many more.

E.g. <ENAMEX TYPE = "LOCOMOTIVE"> वंदे भारत एक्सप्रेस</ENAMEX>

English Translation: वंदे भारत एक्सप्रेस(Vande Bharat Express)

2.1.6   Artefact NE

Artefact entity is considered as an object which is shaped by human craft. It includes tool, ammunition, painting, sculpture, cloth, medicine, gem and stone. The tool can be an object that can help the human to do some work such as a hammer, knife, crockery item, vessel, and computer. Ammunition can be a weapon or a bomb that can be used in war. Painting can be an artwork such as modern art. Sculpture can be the name of a sculpture or a statue. The cloth can be a variety of textile such as silk, cotton, rayon. Gem and stone can be a marble stone, granite, diamond and many more. Medicine can be a kind of ayurvedic, allopathic and homeopathic medicine to prevent and cure disease.

E.g. <ENAMEX TYPE = "ARTEFACT"> त्रिफला चूर्ण </ENAMEX>

English Translation: त्रिफला चूर्ण(Triphala Powder)

2.1.7   Entertainment NE

Entertainment entity is considered as an activity which diverts human attention, gives pleasure and amusement. It includes the performance of some kind such as dance, music, drama, sport, and event. Dance can have different forms such as folk dance-bhangra, ballet. Music can have different forms such as western, Indian classical. Drama can have different forms such as theatre art, cinema, and film. Sport can have different forms such as outdoor- cricket, football; indoor- chess. An event can have different forms such as ceremony, show, conference, workshop, symposium, seminar and many more.

E.g.   <ENAMEX   TYPE   =   "ENTERTAINMENT">
अभिज्ञानशाकुन्तलम्</ENAMEX>

English Translation: अभिज्ञानशाकुन्तलम्(Abhijñānashākuntala)

2.1.8   Cuisine NE

Cuisine entity is considered as a variety of food types that are prepared in different manners such as Chinese, South-Indian, North-Indian foods, etc. It includes different dishes, food recipes such as idly, soup, ice-cream and many more.

E.g. <ENAMEX TYPE = "CUISINE"> चपाती</ENAMEX>

English Translation: चपाती(Chapati/Bread)

2.1.9   Organism NE

Organism entity is considered as a living thing which has the ability to act or function independently such as animal, bird, reptile, virus, bacteria, etc. It also includes human organs. Here, the plant is not considered since it is classified as a separate entity.

E.g. <ENAMEX TYPE = "ORGANISM"> छिपकली</ENAMEX>

English Translation: छिपकली(Lizard)

2.1.10   Plant NE

Plant entity is considered as a living thing which has photosynthctic, eukaryotic, multi-cellular organism characteristics of the kingdom Plantae. It contains chloro-plast, has a cellulose cell wall, and lacks the power of locomotion. It includes the name of the herb, medicinal plant, shrub, tree, fruit, flower and many more.

E.g. <ENAMEX TYPE = "PLANT"> नीम का पेड़ </ENAMEX>

English Translation: नीम का पेड़(Neem Tree)

2.1.11   Disease NE

Disease entity is considered as a disordered state or incorrect functionality of an organ, body part, body system which occurs because of genetic factors, develop-mental factors, environmental factors, infection, poison, toxin, deficiency in nutri-tion value or its imbalance, etc. It can cause illness, sickness, ailments such as fever, cancer; and comprises of disease names, symptoms, diagnosis tests, and treatments, etc.

E.g. <ENAMEX TYPE = "DISEASE"> कुष्ठरोग </ENAMEX>

English Translation: कुष्ठरोग(Leprosy)

2.2   NUMEX Tag

NUMEX[7] entity for the Hindi language includes various numeric expressions based NEs such as count, distance, money, quantity NEs. All such NEs are listed here one by one.

2.2.1   Count NE

Count entity indicates the number or count of item or article or thing etc.
　　For example, <NUMEX TYPE = "COUNT"> दस आम </NUMEX>
　　English Translation: दस आम(10 Mangoes)

2.2.2   Distance NE

Distance entity indicates distance measures such as- miles, kilometre etc.
　　For example, <NUMEX TYPE = "DISTANCE"> 2.5 मील </NUMEX>
　　English Translation: 2.5 मील(2.5 Miles)

2.2.3   Money NE

Money NE: Money entity indicates different units of money.
　　For example, <NUMEX TYPE = "MONEY"> पाँच सौ रुपये </NUMEX>
　　English Translation: पाँच सौ रुपये(Five Hundred Rupees)

2.2.4   Quantity NE

Quantity entity indicates a measure of volume, weight, etc. It also includes expression that conveys some quantity such as- percentage, little, some, etc.
　　For example, <NUMEX TYPE = "QUANTITY"> २ लिटर </NUMEX>
　　English Translation: २ लिटर(2 l)

2.3   TIMEX Tag

TIMEX entity for the Hindi language includes various time-related expressions based NEs such as time, date, day, period NEs. All such NEs are listed here one by one.

---

[7]ltrc.iiit.ac.in/iasnlp2014/slides/lecture/sobha-ner.ppt.

2.3.1 Time NE

Time entity refers to the expression of time such as hour, minute and second.
For example, <TIMEX TYPE = "TIME"> सुबह पांच बजे </TIMEX>
English Translation: सुबह पांच बजे(At Five in the Morning)

2.3.2 Date NE

Date entity refers to the expression of date in different forms such as month, date and year.
For example, <TIMEX TYPE = "DATE"> 15 अगस्त 1947 </TIMEX>
English Translation: 15 अगस्त 1947 (15 August 1947)

2.3.3 Day NE

Day entity refers to the expression of the day such as some special day, days that are weekly, fortnightly, quarterly, biennial, etc.
For example, <TIMEX TYPE = "DAY"> स्वतंत्रता दिवस </TIMEX>
English Translation: स्वतंत्रता दिवस(Independence Day)

2.3.4 Period NE

Period entity refers to the expression of the duration of time, time interval, or time period, etc.
For example, <TIMEX TYPE = "PERIOD"> चौदह साल </TIMEX>
English Translation: चौदह साल(14 Years)

2.4 Other-Than-NE

Other-than-NE considers those names which are mentioned in datasets such as the International Joint Conference on Natural Language Processing (IJCNLP). However, these NEs are fine-grained entity types which are clubbed within certain other NEs to result in coarse-grained NE [15] categorization for an efficient NER system. For example, NEs- abbreviation, brand, designation, a technical term, title-person, and title-object are grouped under the other than NEs.

2.5   Newfangled NEs

Among all the above stated NE types, most of the authors have published work mainly for a person, location, and organization NEs. Recently, authors Jain et al. [43], Jain and Arora [41, 42] have published work on Hindi NER for disease, symptom, consumable, organization and person NEs in the health care domain. Since the health care industry is emerging as one of the largest industries in the world which has a direct impact on the quality of everyone's life and forms an enormous part of our country economy. To work in this direction, the chosen health NEs are related to diagnosis, treatment, and prevention of disease; and is delivered by care providers such as medical practitioners, nursing, pharmacy, and community health workers. For this, Hindi Health Data (HHD) corpus is taken into consideration which is available at the Kaggle dataset[8] and contains more than 105,500 words. The HHD corpus is a quite beneficial resource for the health-based Natural Language Processing (NLP) research.

3   State-of-the-Art for NER in Hindi

This section presents a holistic analysis of the Hindi NER while integrating almost all the major studies in this field. The relevant research articles include full-length articles, survey papers, dissertation, thesis, Hindi NER guidelines that are examined from the six rationale aspects. These Hindi NER research aspects are- datasets, techniques, training corpora, named entities, gazetteers and evaluation measures. A comprehensive review of the rationale aspects is presented here.

3.1   Hindi NER Datasets

The Hindi NER datasets include Message Understanding Conference (MUC) dataset, Translingual Information Detection, Extraction, and Summarization (TIDES) dataset, NLPAI, Shallow Parsing for South Asian Languages (SPSAL) dataset, IJC-NLP NER Shared Task for South and South East Asian Languages (NERSSEAL) dataset, Forum for Information Retrieval Evaluation (FIRE) dataset, International Conference on Natural Language Processing (ICON) dataset, Central Institute of Indian Languages (CIIL) dataset, Dainik Jagaran- popular Hindi newspaper, Gyaan Nidhi corpus- collection of various books in Hindi, web source- tourism domain, HHD corpus- health domain, and varied sources- topics include social sciences, biological sciences, financial articles, religion but not a news corpus. Apart from the above corpora, several other corpora are also encountered such as Language Technologies Research Centre (LTRC) IIIT Hyderabad, tweet corpora- health tweets,

---

[8]https://www.kaggle.com/aijain/hindi-health-dataset.

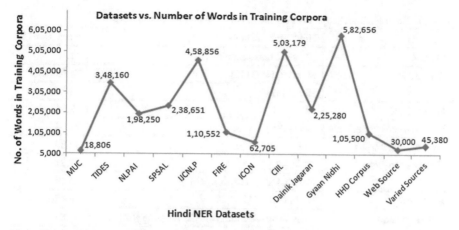

Fig. 2 Hindi NER datasets versus number of words in training corpora

Entity extraction in Social Media text track for Indian Languages (ESM-IL), Code Mix Entity Extraction for Indian Languages (CMEE-IL). It is noted that the NER in Hindi is performed mainly upon these datasets. Figure 2 shows Hindi NER datasets along with the distribution of a total number of words in each of them. Among them, Gyaan Nidhi corpus comprises the highest number of words in the training corpus, followed by CIIL corpus and others.

3.2 Hindi NER Techniques

Several Hindi NER techniques are worked upon by the researchers' over varied datasets w.r.to F-measure that are detailed in Table 1.

In the above table, few techniques result in more than one F-measure values. This is due to certain variant factors such as choice of NE types, distinguished feature sets, gazetteer lists, and datasets that are taken care of by the researchers. To exemplify the same, both [14, 15] consider the CRF technique over the IJCNLP-08 NERSSEAL shared task dataset. Certain features are common to them such as context word, word prefix, word suffix, NE information, and digit information. Also, both contain gazetteers such as first name, middle name, last name, measurement expression, weekdays and month name. However, [14, 15] work for different NE types and few distinguishing features which give different F-measures (Table 2). Other NER techniques and their F-measures as in Table 1 can be justified on the same lines.

Table 1 Hindi NER techniques over datasets are evaluated w.r.to F-measure

Datasets	Hindi NER techniques	F-measure (%)
MUC-6	EM style bootstrapping [9]	41.70
SPSAL 2007	MaxEnt, HMM [3]	71.95
	HMM [12]	78.35
NLPAI 2007	CRF [30]	58.85
BBC and EMI documents	CRF [52]	71.50
Gyaan Nidhi corpus	CLGIN [33]	72.30
Hindi newspapers	ARM [40]	77.81
	Rule-based, list lookup [46]	95.77
Health tweets	HAL, CRF [41]	40.87
Dainik Jagaran	CRF, word clustering, cluster merging [62]	85.83
	MaxEnt, gazetteers, context patterns [67]	81.52
	MaxEnt, gazetteers [69]	81.12
	MaxEnt, word clustering, word selection [70]	79.85
	MaxEnt, semi-supervised learning [72]	78.64
	SVM [73]	83.56
HHD Corpus	HAL, CRF [42]	89.14
	OntoHindi NER [43]	69.33
LTRC Corpus	Bi-directional RNN, LSTM [1]	77.48
ICON 2013	Memory-based, backward elimination [76]	78.37
	HMM [29]	75.20
IJCNLP-08 NERSSEAL	HMM [7]	87.14
	MaxEnt [13]	82.66
	CRF [14]	36.75
	CRF [15]	78.29
	MaxEnt, CRF, SVM [16]	92.98
	SVM [17]	77.17
	GA based classifier ensemble [18]	86.03
	GA-weighted vote based classifier ensemble [19]	72.60
	SVM [20]	80.21
	MOO [21]	92.80
	GA based classifier ensemble- 3 classifiers [22]	92.20
	MOO- MaxEnt, SVM, CRF [23]	93.20
	SVM [24]	89.81
	Ensemble learning [25]	87.86
	Ensemble-based active learning [26]	88.50
	CRF, heuristic rules [28]	50.06
	MaxEnt [37]	82.66

(continued)

Table 1 (continued)

Datasets	Hindi NER techniques	F-measure (%)
	GA [38]	80.46
	HMM, CRF [51]	46.84
	MaxEnt, rules, gazetteers [66]	65.13
	MaxEnt, heuristic rules, context patterns, bootstrapping [71]	96.67
	MOO- 7 classifiers [75]	94.66
	Differential evolution [85]	88.09
	CRF, MaxEnt, rule-based [89]	80.82
FIRE 2010	Bisecting K-means clustering [50]	76.20
FIRE 2013	CRF [83]	96.00
FIRE 2015	CRF [53]	57.59
Twitter	CRF, LSTM [87]	72.06
	DT, CRF, LSTM [88]	95.00
CMEE-IL	SVM [10]	54.51
ESM-IL	CRF [59]	57.55
	CRF [65]	61.61
Web sources	Phonetic matching [58]	65.23
	CRF [32]	64.42
	CRF [84]	45.48
Unknown sources	Rule-based heuristics, HMM [6]	94.61
	CRF, SVM [48]	47.00
	MaxEnt [49]	79.17
	CRF [82]	70.45
	WARMR, TILDE [60]	–
	KNN, HMM [77]	–

Table 2 F-measure for varying NE types and features on CRF technique over IJCNLP-08

Reference	NE Types	Distinguished features	F-measure (%)
Ekbal et al. [14]—CRF	12-person, location, organization, abbreviation, brand, title-person, title-object, time, number, measure, designation, term	word frequency, rare word	36.75
Ekbal and Bandyopadhyay [15]—CRF	5-person, location, organization, miscellaneous, and other-than-NE	first word, infrequent word, word length, POS information	78.29

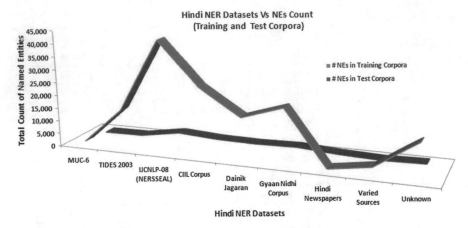

Fig. 3 Hindi NER datasets versus named entities count

3.3 Dataset Versus NE Count

Hindi NER datasets contain training and test corpora having a certain number of named entities. The total count of the number of words as NEs which occur in both training and test corpora are shown in Fig. 3 for the Hindi NER datasets. For example, IJCNLP-08 has the highest number of NEs count in training (43,021) and test (3,005) while CIIL corpus has a number of NEs count in training (26,432) and test (2,022) respectively. TIDES 2003 has a number of NEs count in training (15,063) and test (476) while Hindi Newspapers- Punjab Kesari, Navbhart Times, and Hindustan has a number of NEs count in training (1,030) and test (687) respectively. In addition, some unknown datasets (missing in literature) have a number of NEs count in training (13,767) and test (892), etc.

3.4 NE Tagset Versus NE Count

The hierarchical NE tagset for Hindi NER versus count of each NE type is shown in Fig. 4. Here, the count of NE determines the total number of full-length articles which takes that particular NE type into consideration, and its value is represented as follows. ENAMEX tag comprises of person (54), location (54), organization (53), artefact (8), disease (5), symptom (2), consumable (3), entertainment (10), facility (4), locomotive (10), organism (6), plant (6), and cuisine (4). NUMEX tag comprises of count (4), distance (1), money (6), and quantity (6). TIMEX tag comprises of time (26), date (18), day (4), and period (4). Other-than-NE (12) contains abbreviation, brand, designation, a technical term, title-object, title-person NEs.

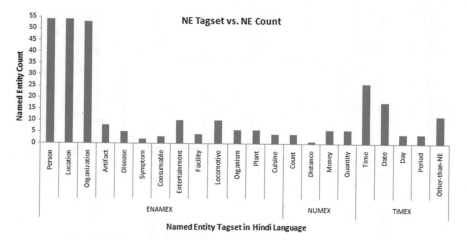

Fig. 4 Hindi NE tagset versus named entities count

3.5 Hindi NER Gazetteers

Gazetteer lists are entities based dictionaries which are important for performing NER effectively [43, 68] as they support non-local model to resolve multiple names of the same entity. Such lists are neither dependent upon previously discovered tokens nor on annotations. Each of these list intakes raw textual input and finds a match based upon its information. Table 3 shows a list of gazetteers which are used by researchers in their Hindi NER related work. For example, person gazetteer comprises of designation, title-person, first name, middle name, last name or surname. Location list comprises of common location, location name, etc. Organization list comprises of the organization name, organization end words. However, certain gazetteers have no further sub-divisions- Jain and Arora [42] have discussed person, disease, symptom, and consumable gazetteers. Ekbal and Bandyopadhyay [15] have discussed month names, weekdays, function words, and measurement expressions gazetteers.

3.6 Datasets Versus Evaluation Measures

The three standard evaluation metrics for Hindi NER- precision, recall, and F-score are taken care of w.r.to different datasets (Fig. 5). For example, IJCNLP-08 has the highest F-score, of 96.67% along with precision as 97.73%, and recall as 95.64% which is followed by FIRE 2013 with F-value as 96.00%, and others.

Table 3 Gazetteers in Hindi NER

Reference	Person name	First name	Middle name	Last name	Person prefix	Person suffix	Designation	Location	Organization name	Organization suffix	Disease
Biswas et al. [3]		Y	Y	Y							
Cucerzan and Yarowsky [9]		Y		Y				Y			
Devi et al. [10]	Y							Y	Y		
Ekbal and Bandyopadhyay [13]		Y	Y	Y							
Ekbal et al. [14]		Y	Y	Y							
Ekbal and Bandyopadhyay [15]		Y	Y	Y							
Ekbal and Bandyopadhyay [17]	Y							Y	Y		
Ekbal and Saha [18]		Y	Y	Y							
Ekbal and Saha [23]		Y	Y	Y							

(continued)

Table 3 (continued)

Reference	Person name	First name	Middle name	Last name	Person prefix	Person suffix	Designation	Location	Organization name	Organization suffix	Disease
Gali et al. [28]	Y										
Gupta and Bhattacharyya [33]					Y	Y		Y	Y		
Hasanuzzaman et al. [37]		Y	Y	Y							
Jain and Arora [41]	Y								Y		Y
Jain and Arora [42]	Y										Y
Jain et al. [43]	Y										Y
Saha et al. [66]		Y	Y	Y				Y			
Saha et al. [67]		Y	Y	Y	Y			Y	Y	Y	
Saha et al. [68]		Y		Y				Y			
Saha et al. [69]		Y	Y	Y				Y		Y	
Saha et al. [71]					Y	Y	Y	Y		Y	
Saha et al. [72]							Y	Y		Y	
Sarkar and Shaw [76]	Y							Y	Y		
Sharma and Goyal [82]	Y				Y			Y	Y		

(continued)

Table 3 (continued)

Reference	Consumable	Symptom	Entertainment	Function words	Season name	Month name	Week day	Time expression	Numerals	Measurement
Biswas et al. [3]						Y				Y
Cucerzan and Yarowsky [9]										
Devi et al. [10]			Y							
Ekbal and Bandyopadhyay [13]				Y		Y	Y			Y
Ekbal et al. [14]						Y	Y			Y
Ekbal and Bandyopadhyay [15]				Y		Y	Y			Y
Ekbal and Bandyopadhyay [17]										
Ekbal and Saha [18]				Y		Y	Y			Y
Ekbal and Saha [23]				Y		Y	Y			Y
Gali et al. [28]								Y	Y	Y
Gupta and Bhattacharyya [33]										
Hasanuzzaman et al. [37]				Y		Y	Y			Y
Jain and Arora [41]	Y									
Jain and Arora [42]	Y	Y								
Jain et al. [43]	Y	Y								
Saha et al. [66]						Y	Y			
Saha et al. [67]						Y	Y			
Saha et al. [68]										

(continued)

Table 3 (continued)

Reference	Consumable	Symptom	Entertainment	Function words	Season name	Month name	Week day	Time expression	Numerals	Measurement
Saha et al. [69]						Y	Y			
Saha et al. [71]					Y	Y	Y			Y
Saha et al. [72]										
Sarkar and Shaw [76]				Y		Y	Y			Y
Sharma and Goyal [82]										

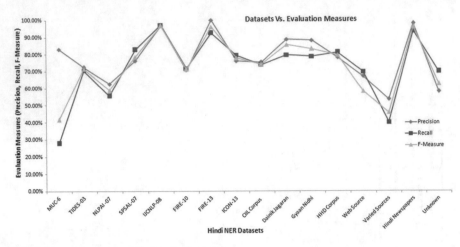

Fig. 5 Datasets versus evaluation measures

4 Contemporary Challenges in Hindi NER

Named Entity Recognition in the Hindi language is much more difficult and challenging as compared to the English. It is due to the wide range of problems that persist in the Hindi NER [45, 61, 78, 86] some of them are discussed here.

- The Hindi language lacks capitalization information which is an important cue for the named entity recognition in the English.
- Although Hindi has a very old and rich literary history, its technological development has a recent origin.
- Hindi names are more diverse and mostly appear as common nouns in the language.
- Hindi is a resource-constrained language- annotated corpus, dictionary, morphological analyzer, part-of-speech tagger and others in the required measure.
- Hindi is a relatively free-word order and highly inflectional language in nature.
- Although gazetteers are easily available for the English, they aren't readily available for the Hindi.
- The Hindi language lacks standardization and spelling of names which result in NE ambiguities.

In addition to the general Hindi NER problems, there are several other domain-specific challenges; some of them are discussed here for the health domain [42].

- NEs e.g. कुटकी चिरौता(Kutki Chirauta) is a consumable NE with rare occurrence within HHD corpus.
- Lack of standardization within abbreviations e.g. Dr. is an abbreviation of Doctor which is represented in Hindi as डॉ, डॉ, डा.
- Postposition marker such as में(in) is present or not with context e.g. सिर में दर्द, सिर दर्दboth represent (Headache).

- Variation in semantic information e.g. मधुमेह, इक्षुप्रमेह, डायबीटीज़ all represent disease NE- (Diabetes).
- Hindi doesn't use the capitalization concept e.g. Acquired Immune Deficiency Syndrome (AIDS) is represented in Hindi as अक्वायर्ड इम्युनोडेफिशिएंसी सिंड्रोम (एड्स).
- The ambiguity between NEs e.g. बुखार (Fever) and खांसी (Cough) can be disease NE or symptom NE.

5 Conclusion

Named entity recognition serves as a vital area of NLP as it has several useful applications. This chapter presents a holistic analysis of NER in the Hindi language while assembling major research works that are done so far in this regard. Major research articles on Hindi NER that are published from 1999 to 2019 are comprehensively reviewed. These articles are examined from six rationale aspects- datasets, techniques, training corpora, named entities, gazetteer lists, and evaluation measures. On unfolding these Hindi NER aspects, researchers gain a better insight into existing problems and proposed solutions for Hindi NE identification and classification. It is highlighted that the widely used Hindi NER datasets are- IJCNLP, Gyaan Nidhi, CIIL, TIDES, SPSAL, Dainik Jagaran, NLPAI, FIRE, ICON, and HHD. Among the NER techniques, a hybrid approach which combines MaxEnt, heuristic rules and bootstrapping gives the best result having F-measure as 96.67% over IJCNLP dataset. Also, the IJCNLP-08 has the highest number of NEs count in the training dataset (43,021 NEs) and in test dataset (3,005 NEs) with the most extensively used NEs as a person, location, and organization. In addition, several other domains specific NEs such as disease, symptom and consumable NEs have emerged for the health corpus with competitive F-measure.

Future Research Directions: Despite the importance of Hindi language, only a few language processing resources are available for researchers such as gazetteers, POS taggers, Hindi WordNet and others. In this regard, there is an abundant research opportunity to fulfill the demand for comprehensive investigation in the Hindi language, especially for the Hindi NER system. Such a task is even more challenging due to lack of appropriate tools and morphological structure of the Hindi. The advancement of Hindi processing tools such as NLP toolkit, well designed NER systems, extensive gazetteers, huge annotated corpora, and computational lexicons are some of the open areas for the research. These tools can facilitate researchers to apply existing approaches in the NER task, as well as many other areas of the Hindi language. Language independent and language dependent feature selection aspect can be investigated. In addition, efforts can be directed towards semi-supervised learning to reduce annotation task, and to provide robust performance for the NER task. Another challenge is to devise NER related models for Twitter and other micro-blogs to handle linguistically challenging contexts over there. While keeping in mind the importance of NER and its applications, there is dire attention for NLP researchers to explore

Hindi Named Entity Recognition in leaps and bounds. Other Indian languages- Bengali, Tamil, Oriya, Urdu, and Telugu also need to be examined.

References

1. Athavale V, Bharadwaj S, Pamecha M, Prabhu A, Shrivastava M (2016) Towards deep learning in Hindi NER: an approach to tackle the labelled data scarcity. arXiv:1610.09756
2. Balyan R (2002) Dealing with Hinglish named entities in English corpora. In: Speech & natural language processing lab, CDAC, Noida, India
3. Biswas S, Mishra MK, Acharya S, Mohanty S (2010) A two stage language independent named entity recognition for Indian Languages. Int J Comput Sci Inf Technol (IJCSIT) 1(4):285–289
4. Bontcheva K, Derczynski L, Roberts I (2017) Crowdsourcing named entity recognition and entity linking corpora. In: Handbook of linguistic annotation. Springer, pp 875–892
5. Chinchor N, Robinson P (1997) MUC-7 named entity task definition. In: Seventh conference on message understanding, vol 29, pp 1–21
6. Chopra D, Jahan N, Morwal S (2012) Hindi named entity recognition by aggregating rule based heuristics and Hidden Markov model. Int J Inf 2(6):43–52
7. Chopra D, Joshi N, Mathur I (2016) Named entity recognition in Hindi using Hidden Markov model. In: 2nd International conference on computational intelligence & communication technology (CICT), pp 581–586. IEEE
8. Cohen WW, Sarawagi S (2004) Exploiting dictionaries in named entity extraction: combining Semi-Markov extraction processes and data integration methods. In: 10th ACM SIGKDD international conference on knowledge discovery and data mining. ACM, pp 89–98
9. Cucerzan S, Yarowsky D (1999) Language independent named entity recognition combining morphological and contextual evidence. In: 1999 joint SIGDAT conference on empirical methods in natural language processing and very large corpora, pp 90–99
10. Devi RG, Veena PV, Kumar, A. M., Soman, K. P.: AMRITA-CEN@ FIRE 2016: Code-mix entity extraction for Hindi-English and Tamil-English tweets. In: CEUR workshop proceedings, vol 1737, pp 304–308
11. Durrett G, Klein D (2014) A joint model for entity analysis: coreference, typing, and linking. In: Transactions of the association for computational linguistics, vol 2, pp 477–490
12. Ekbal A, Bandyopadhyay S (2007) A Hidden Markov model based named entity recognition system: Bengali and Hindi as case studies. In: International conference on pattern recognition and machine intelligence. Springer, pp 545–552
13. Ekbal A, Bandyopadhyay S (2008) Named entity recognition in Indian languages using maximum entropy approach. Int J Comput Process Lang 21(03):205–237
14. Ekbal A, Haque R, Das A, Poka V, Bandyopadhyay S (2008) Language independent named entity recognition in Indian languages. In: IJCNLP-08 workshop on named entity recognition for South and South East Asian Languages, pp 33–40
15. Ekbal A, Bandyopadhyay S (2009) A Conditional random field approach for named entity recognition in bengali and Hindi. Linguist Issues Lang Technol 2(1):1–44
16. Ekbal A, Bandyopadhyay S (2009) A multiengine NER system with context pattern learning and post-processing improves system performance. Int J Comput Process Lang 22(02n03):171–204
17. Ekbal A, Bandyopadhyay S (2010) AEkbalSBandyopadhyay2010Named entity recognition using support vector machine: a language independent approach. Int J Electr Comput Syst Eng 4(2):155–170
18. Ekbal A, Saha S (2010) Classifier ensemble selection using genetic algorithm for named entity recognition. Res Lang Comput 8(1):73–99
19. Ekbal A, Saha S (2010) Weighted vote based classifier ensemble selection using genetic algorithm for named entity recognition. In: International conference on application of natural language to information systems. Springer, pp 256–267

20. Ekbal A, Bandyopadhyay S (2011) Named entity recognition in Bengali and Hindi using support vector Machine. Lingvist Investig 34(1):35–67
21. Ekbal A, Saha S (2011) A Multiobjective simulated annealing approach for classifier ensemble: named entity recognition in Indian languages as case studies. Expert Syst Appl 38(12):14760–14772
22. Ekbal A, Saha S (2011) Weighted vote-based classifier ensemble for named entity recognition: a genetic algorithm-based approach. ACM Trans Asian Lang Inf Process (TALIP) 10(2):1–37
23. Ekbal A, Saha S (2012) Multiobjective optimization for classifier ensemble and feature selection: an application to named entity recognition. Int J Doc Anal Recognit (IJDAR) 15(2):143–166
24. Ekbal A, Saha S, Singh D (2012) Active machine learning technique for named entity recognition. In: International conference on advances in computing, communications and informatics. ACM, pp 180–186
25. Ekbal A, Saha S, Singh D (2012) Ensemble based active annotation for named entity recognition. In: 3rd international conference on emerging applications of information technology (EAIT). IEEE, pp 331–334
26. Ekbal A, Saha S, Sikdar UK (2016) On active annotation for named entity recognition. Int J Mach Learn Cybern 7(4):623–640
27. Etzioni O, Cafarella M, Downey D, Popescu AM, Shaked T, Soderland S, Weld DS, Yates A (2005) Unsupervised named-entity extraction from the web: an experimental study. Artif Intell 165(1):91–134
28. Gali K, Surana H, Vaidya A, Shishtla P, Sharma DM (2008) Aggregating machine learning and rule based heuristics for named entity recognition. In: IJCNLP-08 workshop on named entity recognition for South and South East Asian Languages, pp 25–32
29. Gayen V, Sarkar K (2014) An HMM based named entity recognition system for Indian languages: the JU system at ICON 2013. arXiv:1405.7397 (2014)
30. Goyal A (2008) Named entity recognition for South Asian Languages. In: IJCNLP-08 workshop on named entity recognition for South and South East Asian Languages, pp 89–96
31. Goyal A, Gupta V, Kumar M (2018) Recent named entity recognition and classification techniques: a systematic review. Comput Sci Rev 29:21–43
32. Gupta PK, Arora S (2009) An approach for named entity recognition system for Hindi: an experimental study. In: ASCNT–2009, CDAC, Noida, India, pp 103–108
33. Gupta S, Bhattacharyya P (2010) Think globally, apply locally: using distributional characteristics for Hindi named entity identification. In: 2010 Named entities workshop, association for computational linguistics, pp 116–125
34. Gupta JP, Tayal DK, Gupta A (2011) A TENGRAM method based part-of-speech tagging of multi-category words in Hindi language. Expert Syst Appl 38(12):15084–15093
35. Gupta V, Lehal GS (2011) Named entity recognition for punjabi language text summarization. Int J Comput Appl 33(3):28–32
36. Gupta, V.: Hybrid multilingual key terms extraction system for Hindi and Punjabi text. In: Progress in systems engineering. Springer, pp 715–718
37. Hasanuzzaman M, Ekbal A, Bandyopadhyay S (2009) Maximum entropy approach for named entity recognition in Bengali and Hindi. Int J Recent Trends Eng 1(1):408–412
38. Hasanuzzaman M, Saha S, Ekbal A (2010) Feature subset selection using genetic algorithm for named entity recognition. In: 24th Pacific Asia conference on language, information and computation, pp 153–162
39. Hiremath P, Shambhavi BR (2014) Approaches to named entity recognition in Indian languages: a study. Int J Eng Adv Technol (IJEAT), 3(6):191–194, ISSN:2249-8958
40. Jain A, Yadav D, Tayal DK (2014) NER for Hindi language using association rules. In: International conference on data mining and intelligent computing (ICDMIC). IEEE, pp 1–5
41. Jain A, Arora A (2018) Named entity system for tweets in Hindi Language. Int J Intell Inf Technol (IJIIT), 14(4):55–76 (IGI Global)
42. Jain A, Arora A (2018) Named entity recognition in hindi using hyperspace Analogue to Language and conditional random field. Pertanika J Sci Technol 26(4):1801–1822

43. Jain A, Tayal DK, Arora A (2018) OntoHindi NER—An ontology based novel approach for Hindi named entity recognition. Int J Artif Intell 16(2):106–135
44. Kale S, Govilkar S (2017) Survey of named entity recognition techniques for various indian regional languages. Int J Comput Appl 164(4)
45. Kaur D, Gupta V (2010) A survey of named entity recognition in English and other Indian languages. Int J Comput Sci Issues (IJCSI) 7(6):239–245
46. Kaur Y, Kaur ER (2015) Named entity recognition system for Hindi language using combination of rule based approach and list look up approach. Int J Sci Res Manage (IJSRM) 3(3):2300–2306
47. Khalid MA, Jijkoun V, De Rijke M (2008) The impact of named entity normalization on information retrieval for question answering. In: European conference on information retrieval. Springer, pp 705–710
48. Krishnarao AA, Gahlot H, Srinet A, Kushwaha DS (2009) A comparative study of named entity recognition for Hindi Using sequential learning algorithms. In: International advance computing conference (IACC). IEEE, pp 1164–1169
49. Kumar N, Bhattacharyya P (2006) Named entity recognition in Hindi using MEMM. Technical Report, IIT Mumbai
50. Kumar NK, Santosh GSK, Varma V (2011) A language-independent approach to identify the named entities in under-resourced languages and clustering multilingual documents. In: International conference of the cross-language evaluation forum for European languages. Springer, pp 74–82
51. Kumar P, Kiran RV (2008) A hybrid named entity recognition system for South Asian Languages. In: IJCNLP-08 workshop on NER for South and South East Asian Languages, Hyderabad, India, pp 83–88
52. Li W, McCallum A (2003) Rapid development of Hindi named entity recognition using conditional random fields and feature induction. ACM Trans Asian Lang Inf Process (TALIP) 2(3):290–294
53. Mandalia C, Rahil MM, Raval M, Modha S (2015) Entity extraction from social media text Indian languages (ESM-IL). In: FIRE workshops, pp 100–102
54. Maynard D, Tablan V, Ursu C, Cunningham H, Wilks Y (2001) Named entity recognition from diverse text types. In: Conference on recent advances in natural language processing, Tzigov Chark, Bulgaria, pp 257–274
55. Moro A, Raganato A, Navigli R (2014) Entity linking meets word sense disambiguation: a unified approach. Trans Assoc Comput Linguist 2:231–244
56. Nadeau D, Sekine S (2007) A survey of named entity recognition and classification. Lingvist Investig 30(1):3–26
57. Nanda M (2014) The named entity recognizer framework. Int J Innov Res Adv Eng (IJIRAE), 1(4):104–108. ISSN: 2349-2163
58. Nayan A, Rao BRK, Singh P, Sanyal S, Sanyal R (2008) Named entity recognition for Indian languages. In: IJCNLP-08 Workshop on named entity recognition for South and South East Asian Languages, pp 97–104
59. Pallavi KP, Srividhya K, Victor RRJ, Ramya MM (2015) HITS@ FIRE Task 2015: twitter based named entity recognizer for Indian languages. In: FIRE workshops, pp 81–84
60. Patel A, Ramakrishnan G, Bhattacharya P (2009) Incorporating linguistic expertise using ILP for named entity recognition in data hungry Indian languages. In: International conference on inductive logic programming. Springer, pp 178–185
61. Patil N, Patil AS, Pawar BV (2016) Survey of named entity recognition systems with respect to Indian and Foreign languages. Int J Comput Appl 134(16):21–26
62. Patra R, Saha SK (2019) A novel word clustering and cluster merging technique for named entity recognition. J Intell Syst 28(1):15–30
63. Przybyła P (2016) Boosting question answering by deep entity recognition. arXiv:1605.08675
64. Rahman A, Ng V(2010) Inducing fine-grained semantic classes via hierarchical and collective classification. In: 23rd international conference on computational linguistics (COLING 2010), association for computational linguistics pp 931–939

65. Rao PR, Malarkodi CS, Ram RVS, Devi SL (2015) ESM-IL: entity extraction from social media text for Indian languages@ FIRE 2015-an overview. In: FIRE workshops, pp 74–80
66. Saha SK, Chatterji S, Dandapat S, Sarkar S, Mitra P (2008) A hybrid approach for named entity recognition in Indian languages. In: IJCNLP-08 workshop on NER for South and South East Asian Languages, pp 17–24
67. Saha SK, Sarkar S, Mitra P (2008) A hybrid feature set based maximum entropy Hindi Named entity recognition. In: 3rd International joint conference on natural language processing, vol 1, pp 343–349
68. Saha SK, Sarkar S, Mitra P (2008) Gazetteer Preparation For Named Entity Recognition in Indian languages. In: 6th Workshop on Asian language resources, pp 9–16
69. Saha SK, Ghosh PS, Sarkar S, Mitra P (2008) Named entity recognition in hindi using maximum entropy and transliteration. Polibits 38:33–41
70. Saha SK, Mitra P, Sarkar S (2008) Word clustering and word selection based feature reduction for maxent based Hindi NER. In: ACL-08: HLT, association for computational linguistics, Columbus, Ohio, USA, pp 488–495
71. Saha SK, Sarkar S, Mitra P (2009) Hindi named entity annotation error detection and correction. In: Language forum, vol 35, no 2. Bahri Publications, pp 73–93
72. Saha SK, Mitra P, Sarkar S (2009) A semi-supervised approach for maximum entropy based Hindi named entity recognition. In: International conference on pattern recognition and machine intelligence. Springer, pp 225–230
73. Saha SK, Narayan S, Sarkar S, Mitra P (2010) A composite Kernel for named entity recognition. Pattern Recogn Lett 31(12):1591–1597
74. Saha SK, Mitra P, Sarkar S (2012) A comparative study on feature reduction approaches in Hindi and Bengali named entity recognition. Knowl-Based Syst 27:322–332
75. Saha S, Ekbal A (2013) Combining multiple classifiers using vote based classifier ensemble technique for named entity recognition. Data Knowl Eng 85:15–39
76. Sarkar K, Shaw SK (2017) A memory-based learning approach for named entity recognition in Hindi. J Intell Syst 26(2):301–321
77. Sarkar K (2018) Hindi named entity recognition using system combination. Int J Appl Pattern Recogn 5(1):11–39
78. Sasidhar B, Yohan PM, Babu AV, Govarhan A (2011) A survey on named entity recognition in Indian Languages with particular reference to Telugu. Int J Comput Sci Issues 8(2):438–443
79. Sekine S, Ranchhod E (eds) (2009) Named entities: recognition, classification and use, vol. 19. John Benjamins Publishing
80. Sharma P (2015) Named entity recognition for a resource poor indo-aryan language. PhD Thesis, Department of Computer Science and Engineering School of Engineering, Tezpur University, India
81. Sharma P, Sharma U, Kalita J (2011) Named entity recognition: a survey for the indian languages. In: Parsing in Indian languages, pp 35–39
82. Sharma R, Goyal V (2011) Name entity recognition systems for Hindi using CRF approach. In: International conference on information systems for Indian languages. Springer, pp 31–35
83. Sharnagat R, Bhattacharyya P (2013) Hindi named entity recognizer for NER task of FIRE 2013. In: FIRE-2013
84. Shishtla P, Pingali P, Varma V (2008) A character n-gram based approach for improved recall in Indian language NER. In: IJCNLP-08 workshop on named entity recognition for South and South East Asian Languages pp 67–74
85. Sikdar UK, Ekbal A, Saha S (2012) Differential evolution based feature selection and classifier ensemble for named entity recognition. COLING 2012:2475–2490
86. Singh AK (2008) Named entity recognition for South and South East Asian Languages: taking stock. In: IJCNLP-08 workshop on named entity recognition for South and South East Asian Languages, pp 5–16
87. Singh K, Sen I, Kumaraguru P (2018) Language identification and named entity recognition in Hinglish code mixed tweets. In: ACL 2018 student research workshop, pp 52–58

88. Singh V, Vijay D, Akhtar SS, Shrivastava M (2018) Named entity recognition for Hindi-English code-mixed social media text. In: Seventh named entities workshop, pp 27–35
89. Srivastava S, Sanglikar M, Kothari DC (2011) Named entity recognition system for Hindi language: a hybrid approach. Int J Comput Linguist (IJCL) 2(1):10–23
90. Ugawa A, Tamura A, Ninomiya T, Takamura H, Okumura M (2018) Neural machine translation incorporating named entity. In: 27th international conference on computational linguistics, pp 3240–3250
91. Witten IH, Bray Z, Mahoui M, Teahan WJ (1999) Using language models for generic entity extraction. In: ICML workshop on text mining, pp 1–11
92. Zhu J, Uren V, Motta E (2005) ESpotter: adaptive named entity recognition for web browsing. In: Biennial conference on professional knowledge management/wissens management. Springer, pp 518–529

Data Visualization Techniques, Model and Taxonomy

Shreyans Pathak and Shashwat Pathak

Abstract Data is all around us—even if we realize this or not. From daily weather reports to the fickle changes in the prices of the stock market and even as insignificant as the notification popping up on our cell phones. The data around us is growing exponentially and is expected to grow with a speed unanticipated and hence it becomes very important to store, manage and visualize data. To get a sense of how much data has become relevant to us gets justified by the very fact that almost the amount of data generated since the hundreds of years has been generated in the recent years and it gives us a general sense of how important data has become and it is expected to increase only in the imminent future, and that is why it is important to visualize the data.

Keyword We would like to encourage you to list your keywords within the abstract section

1 Introduction

Data visualization deals with the meaningful and coherent representation of data, which not only makes the whole slew of data more easy to comprehend but also makes the data more palatable to a wide range of population as well. Given the intensity and the amount of data produced every single day the need for data visualization is also increasing day by day. Data visualization incorporates visual representation of data in such a way that the interpretation is lucid and easy to understand. Almost all the professional sections of the society ranging from schools, offices, banks, health care etc. to name a few are relying heavily on data and obviously for representing the data, visualization tools are a necessity. We will see various ways of visualizing data

S. Pathak (✉)
Jaypee Institute of Information and Technology, Sector 62, Noida, UP, India
e-mail: shreyans.pathak@gmail.com

S. Pathak (✉)
Amity School of Engineering and Technology, Sector 125, Noida, UP, India
e-mail: shashwatpathak98@gmail.com

© Springer Nature Switzerland AG 2020
J. Hemanth et al. (eds.), *Data Visualization and Knowledge Engineering*,
Lecture Notes on Data Engineering and Communications Technologies 32,
https://doi.org/10.1007/978-3-030-25797-2_11

as well as the tools that are being widely used for data visualization in the sections that follow [1].

2 Importance of Data Visualization

As mentioned in the sections above, data visualization is graphical representation of data which succinctly provides the same details of information which otherwise to infer from numbers would be comparatively hard to interpret. What really makes this even more important is the fact that it also serves as a huge time saver both in terms of presenting and interpreting. Although the data to be represented might need to be processed prior to the visual representation. Another important aspect associated with data visualization is that apart from the data being easier to digest, the trends can be predicted as well which can considerably help in decision making. Data visualization can also elicit the information or some aspect of data that might normally get overlooked in absence of data visualization tools and that is why scientists and engineers rely heavily on data visualization tools for creating reports to analyze the data better. So owing to the added efficiency and cost saving as well since the time needed to effectively analyze data is cut short to a huge amount adding to the overall advantage.

3 Steps Involved in Data Visualization

We have seen the importance of data visualization and how it is beneficial for representation of data in succinct yet lucid way. The ability of data visualization to simplify and transform the complex information and details of a data in an easier and insightful form makes the use of data visualization even more important, [2] but prior to performing this simplification, we might need to perform few actions, which we commonly term as the 'Pre-processing' of data. There are various steps involved in the same which are worthy of a discussion. On a high level, we can classify the steps involved in the pre-processing as follows [3]:

- Data collection
- Data parsing
- Data filtering
- Data mining
- Data representation.

Now, we can talk about these steps in detail. The number of steps that we follow is variable, in some cases we might need to follow only few steps out of all of these and in some of the cases we might need to go through the complete set of steps in order to reach to the conclusion. We will briefly go through the procedure involved in these steps.

3.1 Data Collection

This is the first and very important part of the whole process. The collection or acquisition of the data is of the primary importance because this step can set the course of action to be followed for rest of the steps as the data collected can be quite complex or it can be as easy as just uploading a file containing the desired raw data. The data collection itself has a large scope and it involves various techniques using which we can collect information for the data visualization. Some of the commonly used techniques employed for data collection are as follows:

Interviews The most basic technique of gathering data is the interview. The interview can be conducted in person or online, the process involves the conversation based on a set of questions using which the responses of the interviewees are gathered and stored. The interviews can be performed in a formal or informal manner aiming at putting forth the questions that are easy to understand and that can elicit the information from the person being interviewed.

Surveys Surveys or online questionnaires are also getting popular as the means of collecting information for the data that we wish to visualize. The surveys also have a predefined set of questions in which the user has to choose the options that the user thinks is the best and the response of each of the user is stored in some file (excel sheet or CSV etc.). Each of the options can be assigned some numerical value which can facilitate in the final analysis, and due to this reason the analysis can be slighter easier as compared to Interviews. Employee satisfaction surveys are much in use nowadays by many businesses and similarly many product companies have also taken the advantage of the pervasive internet connectivity to create online surveys where they can ask the users to rate their experiences.

Observation Data gathering can also be achieved using some observational techniques in which the researcher observes the participants and collects the relevant information from the same. Observations can be of different types namely

– Participant and Non Participant Observation
– Behavioral Observation
– Direct/Indirect Observation
– Covert/Overt Observation.

Each of the techniques pertaining to the observational way of collecting data have their own pros and cons and choosing the one depends on many factors including the complexity of the research being carried out. As far as the observational approach is concerned it gives the researcher the opportunity to create and test real case scenario and also instead extrapolating the results hypothetically the researcher amasses a real chunk of data. However, the data collection can get marred by personal bias of the researcher and this is also prone to fake behaviour from the subjects under observation.

Focus Groups Another mean of collecting data are the focus groups which are essentially group of people with varied demography and the members of the group interact with each other on topics related to politics, some product or some belief and the data is gathered based on the interaction of the participants of the focus group by the means of taking notes by the researcher.

3.2 Data Parsing

After the essential step of collecting the data, the next step is to make sense of the data. This involves converting the raw data gathered in the previous stage into a meaningful form. This might involve addition of delimiters to the collected data or adding different rows and columns to the collected data and label them.

3.3 Data Filtering

Not all the data parsed for further processing might actually be of full use and hence we might need to filter some of the data that we might deem as irrelevant. For instance if we have got the data containing the voting information of the all the regions and the data that we wish to analyze is pertaining to the north region only then rest of the data is out of the scope of the intended field of study and hence can be ignored for the actual analysis.

3.4 Data Mining

Data Mining is a huge discipline in itself [4], it encompasses the extraction of information from huge sets of data and it involves extensive use of various mathematical as well as statistical techniques to unearth the sense out of plethora of data, and thus the name Data mining. Data mining ever evolving and now it has taken advantage of Artificial Intelligence as well [5] for performing analysis of tasks like fraud detection, pattern analysis, marketing just to name a few. Traditionally used on the relational databases, it has now extended to Data Warehouses. We will briefly look into some of the techniques and processes involved in data mining.

Understanding the data This step is basically the validation of the data collected in the steps mentioned above. The data collected from various sources is collated and the relationships between them is established so that the data can be merged for the further processing, this can be a tricky step as the data that has been fetched can be of different type and care should be taken while merging the data. This step can also ensure that whether the data is complete or not, and any data that is missing or is inconsistent must be incorporated at this preliminary step.

Data Representation Finally this is where we can define the form that the data will take, we can choose the data to be represented in the form of a graph with the axes labelled with the corresponding values. This is a crucial step as here we decide on how to represent the data that will be used for further refinement and finally will be able to be represented in the form of a suitable chart or graph that can visualize the data for the end user [6].

Data Preparation In this step, the data is processed further and the main intention at this point is to remove the noise from the data and this takes a considerable amount of time as the some time can be taken for cleaning and smoothing of the data. At this point again in the final analysis if any data is found to be missing or incomplete then that can be added for improving the overall consistency of the data set.

Data Transformation Data Transformation is another step towards the data mining and aims at further preparing the data ready for the final processing. This step involves the following

- Smoothing
  This step is involved in removal of noise from the collected data as mentioned above.
- Aggregation
  This step aims at the collation of the collected data and to apply some aggregate functions to calculate some values if needed.
- Generalization
  This step is mainly concerned with the possibility of modification of any low level detail with any high level information [7], if possible to ease the understanding of the data. For Instance, if we have a data of a population of male and female vote percent in any constituency, then we can replace this with combining the male and female votes in one combined set.
- Normalization
  Usually in the data set we have some constraints or value ranges within which the data normally lies and the aim of performing the normalization is this only. The data set is scaled so that it falls within a definite range. This might require mathematical operations for achieving the normalization [8]. For instance, we might need to normalize the data set so that the data must fall between range 0 to 20 only.

Data Modelling In this step, we develop the mathematical models against which we analyze the data patterns. This is again dependent on many factors

- The most important one being business objective as in most of the cases ultimately the data is getting refined for some business purposes only so the modelling must done in a way that it keeps up with the business objectives [9]. Once the model has been finalized the next step is to run the model over the data and analyze the results and decide that whether the data set actually meets the need. There are two further steps involved in this step

– Evaluation
– Deployment.

As the name suggests, the evaluation is concerned with assessing if the mathematical model is consistent with the business objectives and any further decision on whether to carry on with the developed model or not. This step acts as the making or breaking point. The next step is Deployment—if the model is consistent with the business needs then the data should be made available for the common stakeholders and it should be easy to understand for the end users.

Data Mining Techniques The data mining follows a set of procedures which facilitates the whole data mining process [10], we will go through the brief description of the steps which are mentioned below

– Classification
  As the name suggests, this step is used for pulling the meta-data and in general to classify the data, i.e. to put them in different classes.
– Clustering
  As the name suggests, in this technique we divide the data into different groups that are commonly known as clusters, the intention being to keep the similar data together. This immensely helps us to identify what is the level of similarity between the elements of the data-set.
– Regression
  This technique is used for making the predictions based on the existing set of data and mostly employs probabilistic models that can predict what the plausible outcome might be based on the input values [11].

Types of Data Mining On the high level, there are two main categories of the data mining techniques—namely Supervised and Unsupervised learning techniques. These techniques form the major portion of Artificial Intelligence and are widely used in predictive and classifying algorithms. We will go through them briefly [12].

– Supervised Learning
  The supervised learning is aimed at the prediction and classification tasks and generally begins with a dataset containing some values and the we develop a model that will learn based on these values and can make predictions about the output based on the input parameters. Generally the models that are used in supervised learning are as follows

  • Linear Regression
    Here the output is predicted based on the continuous values of the variables used and generally it is ideal for the tasks that involve predictions based on single or multiple input values. For example rating the value of a car based on some of the input parameters such as engine, make etc.
  • Neural Networks
    This model is inspired by the structure and functioning of the brain itself and that is why it is named so [13]. Just like the brain neurons fire and disseminate the information further, here also units called as neurons are triggered based on the

inputs on whether or not it satiates the threshold value, using which the signal is fired or not. This approach finds extensive usage in pattern recognition and particularly Optical Character Recognition (OCR) has been heavily influenced by neural networks and so does data mining as well.

- K-Nearest Neighbour

  This model is mostly dependent on data, as the name suggests the model categorizes the observation based on past observations and to do so it checks it's K nearest neighbours, where K is a positive integer and puts the current input with the class of the identified neighbour in order to classify an input.

- Unsupervised Learning

  Unsupervised data is concerned with the task of finding the inherent patterns in the data, unlike supervised learning there is no preexisting dataset to glean information from it learns from the patterns itself and employs classification in order to decode the patterns in the input sets [14].

4   Working of Data Visualization

We have seen how visualization can be an indispensable tool for representing and analyzing data, but as mentioned earlier, the data that has to be represented in a visual form has to be processed before we can actually extract some meaningful information from the same because the data being dealt with can be collosal in terms of size and complexity and this is when various tools and software obviate the need of manual human work for data representation. The most common and widely used tools like Microsoft Excel, Microsoft PowerPoint are known to everyone, but now over the course of time many new software and tools have been developed that makes the lives of the data analyst even easier. Another term associated with data visualization is the database, where the data is actually stored in the raw format and these tools extract the data from the database and use the same to create a visual image of the otherwise non representative data [15]. The level of sophistication of these tools also vary and depend on the need of the user. The data can also be pulled from some file, for instance a comma separated value (CSV) file to a similar effect. These tools induce a completely new insight to the data and it's representation, the relations and all the statistical measures of the data like mean, standard deviations, correlations etc. can be visualized with these specialized tools. The data extracted can be represented in a plethora of forms which we will see in the upcoming sections [16].

5   Means of Data Visualization

There are numerous ways in which we can represent the graphical data, apart from the traditional line charts now there are many more different and even more lucid forms of data visualization which we will see one by one [17].

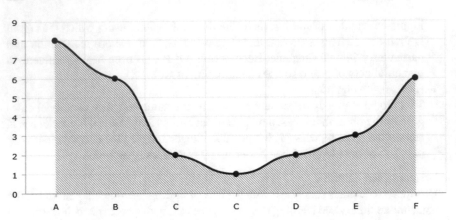

Fig. 1 A simple line chart

5.1 Line Charts

The Line charts depict the relationship in two variables, widely used over the course of time in many fields notably in mathematics, physics and statistics. They are normally used to depict the change in one of the variables compared with the corresponding change in other. So essentially we can represent the relationship between the variables easily in a line chart and this should be the factor while deciding on whether to choose the line chart or not for representing the relationship. The line chart is depicted in the diagram (Fig. 1).

5.2 Bar Charts

Bar charts are also one of the most common way of data representation and visualization. The Bar Charts are commonly used for comparing data of different types of groups or categories. The values are easily visible from the eye although sometimes we might need to use some kind of color coding in order to distinguish the groups. The most common variations to the Bar Charts is the horizontal or vertical orientation, either form can be used as per the requirement. One kind of Bar Chart is represented in the diagram (Fig. 2).

Sometimes, we might need to display different kinds of parameters in the same bar and in order to do that we can stack some other parameters on the same bar, which however, can be lesser intuitive as the starting and the ending values of the parameters have to be figured out manually using the values it the axis. One of such graphs is depicted in the diagram (Fig. 3).

In yet another variation, we can put the line chart and the bar chart together to as depicted in the figure (Fig. 4).

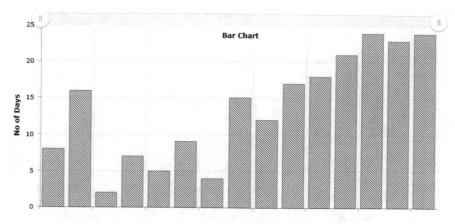

Fig. 2 A simple bar graph

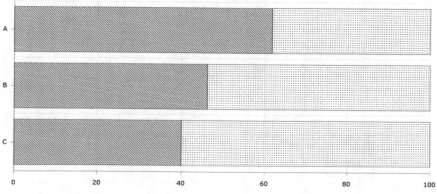

Fig. 3 A stacked bar graph

5.3 Scatter Plots

The scatter plots are also another mean of data representation, also known as x-y plots widely used in statistics due to their ability of representing the correlation among the variable being plotted. The relation between the variables can be represented if they have a relation between themselves. Using the scatter plots can help us visualize the spread of the data or how closely they are related. It serves as an indispensable tool when the amount of data is large, otherwise a line chart or a bar chart can also suffice the requirements. One of the scatter plots is shown in the diagram below (Fig. 5).

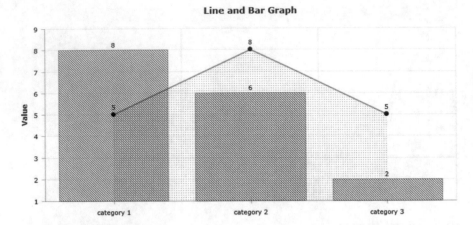

Fig. 4 A line and bar graph

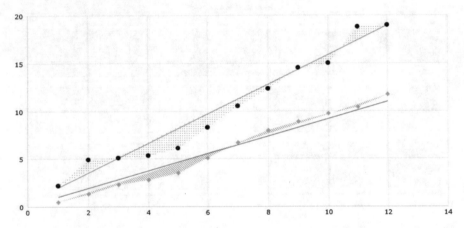

Fig. 5 A scatter plot

5.4 *Pie and Donut Charts*

In the pie and donut charts the data is represented in terms of part of a whole, in the pie chart the data is represented as the slice of the complete 'pie' and donut charts also represent in similar way, just the difference being that they are represented in the form of donut [18]. Although in an unlabeled chart or unless mentioned otherwise we might need to make a guess about the values of the different sectors, so labelling the amount of the values might be a good idea for comprehending the values better (Fig. 6).

Fig. 6 A pie chart

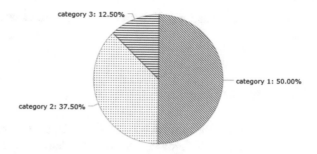

5.5 Other Methods of Data Representation

The different forms of data visualization methods we mentioned so far are not limited till there and as a matter of fact new variants keep on appearing and the existing representations keep on getting revamped. Some other ways of data visualization are as follows [19]

- Box and Whisker plots
- Heat Map
- Gantt Chart
- Histogram
- TreeMap.

6 Data Visualization Tools

There are various tools available that can effectively visualize the data for the end user and offer all level of sophistication as per the users need. Some of the widely used data visualization tools currently available are as follows.

6.1 Tableau

Tableau is one of the most popular and easy to use data visualization tool available and due to this very reason the user base of Tableau has been increasing day by day. The integration of Tableu with various databases give it an edge and provide much flexibility as well. It is easy to use, offers many features using which we can customize and present the data well. Apart from that it is also very easy to use from the mobile devices. This tool is licensed.

6.2 Zoho Reports

This tool also allows the user to create reports and charts ans dashboards. Just like the Tableau it also connects to various databases and the user interface is also very appealing, but sometimes the application might run slow. It has an online version as well.

6.3 Infogram

Infogram allows the users to connect to real time data and has a lots of templates to choose from and also provides tailored templates for users of many professions.

Now apart from the tools offered by these softwares, over the course of time we the development platforms have also created many libraries for the users where they can perform the data visualization according to their needs without any restriction as in the case of many licensed software like Tableau. Now We can discuss the functionalities offered by various programming languages related to data visualization.

6.4 Data Visualization in Python

Python has to offer many great tools and libraries that are dynamic, versatile and at the same time very powerful [20]. Some of the widely used python libraries are listed down below that are used extensively for data visualization.

- Pandas
- Matplotlib
- Plotly
- ggplot
- Seaborn.

The first step here as well is to start with a dataset. There are various datasets that are available online in the form of CSV files that can be freely downloaded and used for the purpose of data visualization.

We will be creating various charts to demonstrate the capabilities and ease that these libraries provide.

Importing data Once the Dataset has been selected, then we can use the *Pandas* library which offers function read csv in which we can pass the dataset file name and the data will be read from the CSV file. We can use the below code. Here we are using the famous Iris dataset which contains the data about the different flower type and sizes, the file here is a CSV file and we will use the same for demonstrating the data import using the pandas library offered in python [21]

```
import pandas as pan
data = pan . read csv-(' iris . csv' )
```

After the execution of the above code, we get the complete data set as (Fig. 7).

Fig. 7 Imported DataSet

	sepal.length	sepal.width	petal.length	petal.width	variety
0	5.1	3.5	1.4	0.2	Setosa
1	4.9	3.0	1.4	0.2	Setosa
2	4.7	3.2	1.3	0.2	Setosa
3	4.6	3.1	1.5	0.2	Setosa
4	5.0	3.6	1.4	0.2	Setosa
5	5.4	3.9	1.7	0.4	Setosa
6	4.6	3.4	1.4	0.3	Setosa
7	5.0	3.4	1.5	0.2	Setosa
8	4.4	2.9	1.4	0.2	Setosa
9	4.9	3.1	1.5	0.1	Setosa
10	5.4	3.7	1.5	0.2	Setosa
11	4.8	3.4	1.6	0.2	Setosa
12	4.8	3.0	1.4	0.1	Setosa
13	4.3	3.0	1.1	0.1	Setosa
14	5.8	4.0	1.2	0.2	Setosa
15	5.7	4.4	1.5	0.4	Setosa
16	5.4	3.9	1.3	0.4	Setosa
17	5.1	3.5	1.4	0.3	Setosa
18	5.7	3.8	1.7	0.3	Setosa
19	5.1	3.8	1.5	0.3	Setosa
20	5.4	3.4	1.7	0.2	Setosa
21	5.1	3.7	1.5	0.4	Setosa
22	4.6	3.6	1.0	0.2	Setosa
23	5.1	3.3	1.7	0.5	Setosa

Scatter Plots Now we can use the *Matplotlib* offered in python to generate the graphs and we can visulize the data using the functionalities offered by matplotlib [22]. We will use the below set of code and the variable *data* that we created above to generate the scatter plot of the dataset.

```
%matplotlib inline
import matpl otl ib . pyplot as matplt
# c r eating the axes
axes = matplt . subplots ()

# Scatter p l o t of the l e n g t h and the width
axes . s c a t t e r ( data [' s e pal . l ength ' ], data [' s e pal . width ' ])

# s e t t i t l e and l a b e l s
axes . set title (' Iris Dataset ' )
axes . set_xlabel (' Length' )
axes . s e t y l a b e l (' Width' )
```

The output of above lines of code is shown in the diagram (Fig. 8).

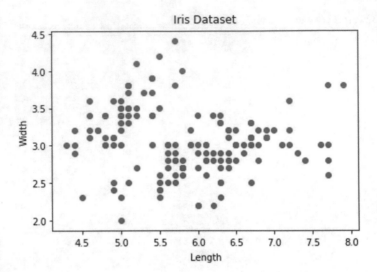

Fig. 8 Dataset scatter plot

Line Chart We can also create a line chart using the same data set, we can use the *plot* method to do the same. We will write the code below to get the line chart.

```
# get columns to p l o t
columns = data . columns . drop ( [ ' v ari e ty ' ] )
# create x data
x data = range ( 0 , data. shape [ 0 ] )
# f i g u r e and axis
axes = pl t . subplots ()
# plotting each column
for c o l in columns :
        axes . pl o t ( x data , data [ c o l ] )
# s e t t i t l e and l egend
axes . s e t t i t l e ( ' Line Chart − Iris ' )

#adding l egends as w e l l
axes . l egend ()
```

After executing the above lines of code we get the output as (Fig. 9).

Histograms Coming to the next data visualization form, we can also create the histograms using the same data set by Matplotlib. The same is demonstrated in the code below

```
# axis c reation
axes = pl t . subplots ()

# p l o t t i n g the histogram
axes . h i s t ( data [ ' s e pal . l ength ' ] )

# s e t t i t l e and l a b e l s
axes . set title ( ' Iris Sepal Length Histogram ' )
axes . set xlabel ( ' Length ' )
axes . s e t y-l a b e l ( ' Frequency ' )
```

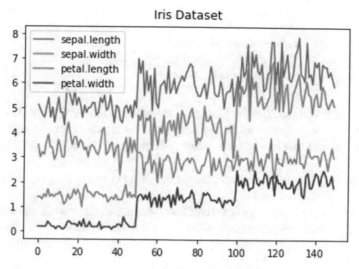

Fig. 9 Line chart plot using matplotlib

The output of the code above will result in a histogram which is plotted as the Sepal length and the frequency (Fig. 10).

Similarly we can plot for the Petal length and rest of the parameters as per our need.

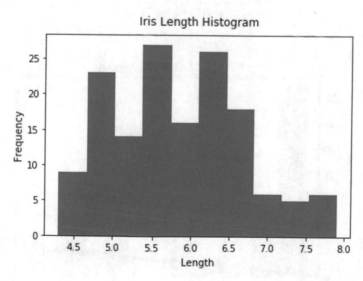

Fig. 10 Histogram of the sepal length

```
# axis c reation
axes = pl t . subplots ()

# plotting the histogram
axes . hist ( data [' pe tal . length ' ])
axes . set title (' Iris Petal Length Histogram ' )
axes . set xlabel (' Length ' )
axes . set y-label (' Frequency ' )
```

Above code will result in the histogram as shown in the figure (Fig. 11).

It is very easy to create multiple histograms all at the same time using the power of pandas, using just a line of code. Same is demonstrated in the diagram below

```
data . pl o t . h i s t ( subplots=True , layout =(2 , 2) , f i g s i z e =(10 , 10) , bins =20)
```

Bar Charts Now moving on to the Bar Charts, we can create the bar chart similarly, which is demonstrated in the lines below (Figs. 12, 13, 14, 15).

```
# Axis
axes = pl t . subplots ()

# counting the varieties
barGraph = data [' v ari e ty ' ] . value counts ()

# Data f or the coordinates
poi nts = barGraph . index
f requency = barGraph . values
# create bar chart
axes . bar ( points , f requency )

# Tit l e o f the chart
axes . set title (' Iris Variety Graph' )
axes . set_xlabel (' Variety ' )
axes . set y-label (' Frequency ' )
```

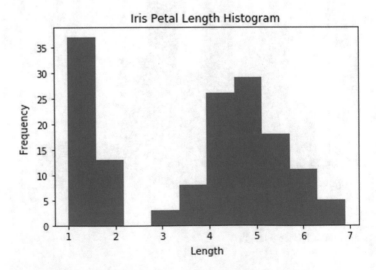

Fig. 11 Histogram of the petal lengths

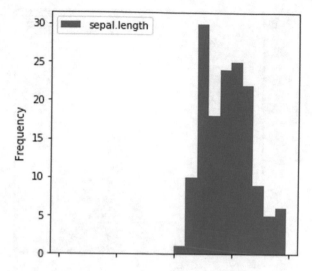

Fig. 12 Histogram of the sepal lengths

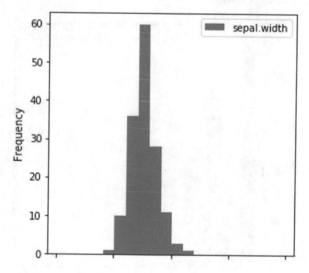

Fig. 13 Histogram of the sepal width

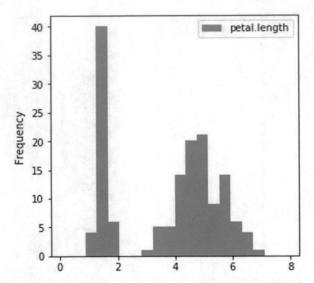

Fig. 14 Histogram of the petal lengths

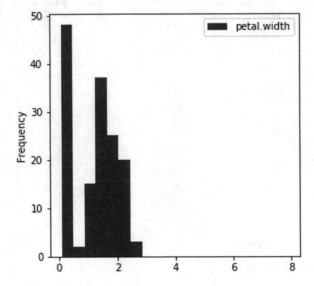

Fig. 15 Histogram of the petal width

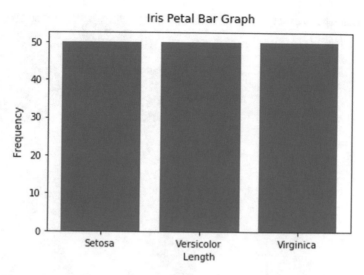

Fig. 16 Bar graph of the flower varieties

On executing the code above, we get the following output (Fig. 16).

HeatMaps Heatmap is another representational way in which the frequencies of the various parameters of the data set is represented in different colors, much like an image captured by a thermal imaging camera in which the graph consists of varying temperatures and the temperatures are differentiated according to the colors. We are using *numpy* here which is another library specializing in array processing

To create such a map we can use the below line of code

```
import numpy as nump
# correlation matrix
cor = data . c o rr()
axes = pl t . subplots ()
# heatmap generation
image = axes . imshow ( cor . values )

# labels
axes . set xticks (nump. arange (len (cor . columns )))
axes . set yticks (nump. arange (len (cor . columns )))
axes . set xticklabels (cor . columns )
axes . set -yticklabels (cor . columns )

# we can s e t the ro t a t i o n and other parameters here
plt . setp (axes . get _xticklabels (),  rotation =45, ha=' right' ,
          rotation_mode=' anchor' )
```

The heatmap generated from the code above is shown in the diagram (Fig. 17).

Fig. 17 HeatMap of the data set

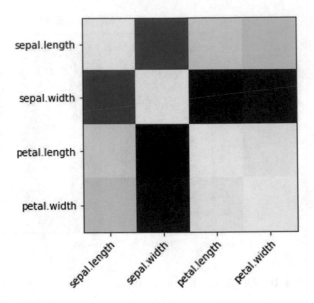

Another great library offered in python is *Seaborn* which also provides great functionalities for plotting and creating the graphs. We can generate the above heatmap with added annotations, which greatly enhances the overall information provided by the graph. This is demonstrated below

```
import seaborn as sns
# firstly create the correlation matrix
corrl = data.corr()
axes = plt.subplots()
# create heatmap
im = axes.imshow(corr.values)

# set labels
axes.set_xticks(np.arange(len(corrl.columns)))
axes.set_yticks(np.arange(len(corrl.columns)))
axes.set_xticklabels(corrl.columns)
axes.set-yticklabels(corrl.columns)

# Rotating thel labels - not a necessary step
plt.setp(ax.get_xticklabels(), rotation=45, ha='right',
        rotation_mode='anchor')

# creating the annotations to be displayed
for j in range(len(corrl.columns)):
    for k in range(len(corrl.columns)):
        text = ax.text(k, j, np.around(corrl.iloc[j, k], decimals=2), ha='cen
```

The output of the above lines of code is as (Fig. 18).

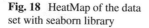

Fig. 18 HeatMap of the data set with seaborn library

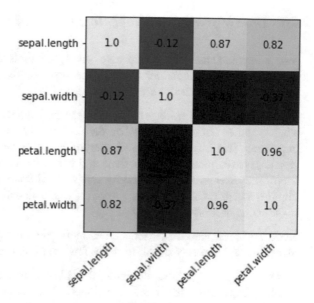

6.5 Data Visualization Tools in R

We saw that we can easily implement data visualization in python thanks to the plethora of libraries available, along with the amount of functionality offered. We can easily chew down massive information to digestible bits owning to the enhanced abilities provided by these visualization tools. But the Data Visualization tools are not limited to python only. We have got visualization tools on other platforms as well that can do the complex task of visualizing the data a breeze. We will discuss some of the popular tools on available in R first [23].

R is a programming language which was developed exclusively for providing statistical computation and graphics functionality and was released in 1993. Since then many packages (or libraries) have been developed that provide powerful data visualization capacities to the end users. Some of the popular Data visualization tools have been listed down below

ggplot2 ggplot2 is one of the most commonly used library in R, using which creating custom plots and graphs becomes very easy. Loaded with built in functions, the tedious task of data visualization becomes very easy when it comes to ggplot2.

Lattice Another great package for data visualization, Lattice is very helpful when it comes to plot multi-variable data. We can create the partitions and tiles that can facilitate to graph different variables on the same chart.

Highcharter Also available in JavaScript, Highcharter provides many built-in functions using which we can create interactive charts of all kinds, the package provides flexibility in creating custom plots and graphs at the same time maintaining the brilliant user interface.

6.6   Data Visualization Tools in JavaScript

Just like Python and R, JavaScript also offers great tools for web developers to create visualizations and animations, some of the most popular libraries available for Data Visualization in JavaScript are enlisted below [24].

D3.js This library is extremely popular and perhaps the most widely used library when it comes not only to visualize data but also to animate and creating graphics in general. This library has almost everything that a web developer might need for creating charts, plots and animations and hence the popularity of D3.js speaks for itself. The extensive API offered by D3.js sets it apart from rest of the JavaScript libraries and as mentioned before—not only data visualization but animation is also in the scope of this huge library.

Recharts Developed for React.js developers, this library uses D3.js for it's implementation and in a way simplifies what few would find as complex in D3.js.

Chart.js Lightweight and efficient—this library provides exclusive charting and graphing capabilities to the web developers. The implementation is also very easy when it comes to Chart.js, it offers some fixed charting options and is very popular for the beginners.

7   Conclusion

So, in this chapter we have seen how important data visualization is for providing a graphical representation of data which otherwise is not that easy to comprehend without applying visualization techniques. We have also seen the amount of pre-processing that has to be done in order to get the data ready for being able to be visualized, starting from the data collection techniques and then cleaning and filtering to final processed data. We also explored how Machine learning and artificial intelligence techniques [25] like supervised and unsupervised learning are being used extensively in the data visualization; providing the most efficient and insightful image of the data under consideration. We also saw the actual implementation of data visualization using a dataset in the most common implementation as of now—Python. Apart from python we also saw how other programming languages like R and JavaScript also have provided us the capability to visualize the data effectively for the end user. So this chapter devoted mainly to data visualization and the techniques demonstrates the ease and insight that visualization provides. The future of data visualization is even more exciting with more powerful tools and algorithms being developed and we can say that the imminent future for data visualization and analytic is very bright indeed.

References

Janert PK (2010) Data analysis with open source tools. O'Reilly Media Inc, CA

Dbler M, Gromann T (2019) Data visualization with python: Create an impact with meaningful data insights using interactive and engaging visuals. Packt Publishing, Birmingham, UK

Fry B (2007) Visualizing data. O'Reilly Media, Inc, CA, pp 264–328

Han J, Kamber M (2011) Data mining: concepts and techniques, 3rd edn. Morgan Kaufmann, US

Bishop CM (2006) Pattern recognition and machine learning (information science and statistics), 1st edn. Springer, Berlin, Germany

Cairo A (2012) The functional art: an introduction to information graphics and visualization. (Voices That Matter), New Riders, San Francisco, US

Steele J (2011) Designing data visualizations: representing informational relationships. O'Reilly Media Inc, CA

Rice JA (2013) Mathematical statistics and data analysis. Cengage Learning, MA, US

Knaflic CN (2015) Storytelling with Data: a data visualization guide for business professionals. Wiley, NJ, US

Hastie T, Tibshirani R, Friedman J (2017) The elements of statistical learning: data mining, inference, and prediction. Springer, Berlin, Germany

Chatterjee S, Hadi AS (2013) Regression analysis by example, 5ed (WSE). Wiley, NJ, US

Manning CD, Raghavan P, Schtze H (2008) Introduction to information retrieval. Cambridge University Press, Cambridge, UK

Fausett LV (1993) Fundamentals of neural networks: architectures, algorithms and applications, 1e. Pearson, London, UK

Harrington P (2012) Machine learning in action, manning publications, New York, US

Rendgen S (2012) Information graphics. TASCHEN, Cologne, Germany

Cairo A (2016) The truthful art: data, charts, and maps for communication. New Riders, San Francisco, US

Steele J, Ilinsky N (2010) Beautiful visualization. O'Reilly Media Inc, California

Evergreen S (2016) Effective data visualization: the right chart for the right data. SAGE Publications Ltd, CA, US

Wilke CO (2019) Fundamentals of data visualization: a primer on making informative and compelling figures. O'Reilly Media Inc, CA, US

McKinney W (2017) Python for data analysis: data wrangling with pandas, numpy, and ipython, pearson education. O'Reilly Media Inc, CA

Chen DY (2018) Pandas for everyone: python data analysis, 1e. Pearson Education, Pearson, London, UK

Nelli F (2018) Python data analytics: with pandas, numpy, and matplotlib, 2nd edn. Apress, New York, US

Wickham H (2017) R for data science: import, tidy, transform, visualize, and model data, pearson education. O'Reilly Media Inc, CA

Dale K (2016) Data visualization with python and javascript: scrape, clean, explore & transform your data. O'Reilly Media Inc, CA

Mller AC, Guido S (2016) Introduction to machine learning with python: a guide for data scientists. O'Reilly Media Inc, CA

Prevalence of Visualization Techniques in Data Mining

Rosy Madaan and Komal Kumar Bhatia

Abstract Due to huge collections of data, exploration and analysis of vast data volumes has become very difficult. For dealing with the flood of information, integration of visualization with data mining can prove to be a great resource. With the development of a large number of information visualization techniques over the last decades, the exploration of large sets of data is well supported. In this chapter, we present a detailed explanation of data mining and visualization techniques. The chapter also discusses how visualization can be applied in real life applications where data needs to be mined as the most important and initial requirement. A detailed explanation of graphical tools and plotting various types of plots for sample datasets using R software is given.

keywords Data mining · Data visualization · KDD · Graphs · Information retrieval · Text mining · Plots

1 Introduction

Data mining [1, 27] is a process of seeking for the identification of patterns which are hidden, valid and possibly helpful in large sets of data. Data Mining is concerned with finding out unsuspected relationships amongst the data which are not previously known. It is a multidisciplinary area that uses AI, machine learning, statistics and database technology. The understanding derived via Data Mining can be applied into various areas like for detection of fraud, in marketing and for scientific discovery etc. Other terms used for data mining are information harvesting, Knowledge extraction, Knowledge discovery, data/pattern analysis etc.

R. Madaan (✉)
Computer Science & Engineering, Manav Rachna International Institute of Research and Studies, Faridabad, India
e-mail: madaan.rosy@gmail.com

K. K. Bhatia (✉)
Computer Engineering, J.C. Bose University of Science and Technology, YMCA, Faridabad, India
e-mail: Komal_bhatia1@rediffmail.com

© Springer Nature Switzerland AG 2020
J. Hemanth et al. (eds.), *Data Visualization and Knowledge Engineering*,
Lecture Notes on Data Engineering and Communications Technologies 32,
https://doi.org/10.1007/978-3-030-25797-2_12

There is nothing new in saying that data mining is really a subject related to different disciplines and it can be explained in numerous ways. It is very difficult to identify all the components from the term **data mining**. The drawing out of gold from big rocks or sand can be referred to as gold mining not as rock or sand mining. Similarly, "knowledge mining from data," can be a more appropriate term than the data mining.

In addition to this, many other terms can be used as they have meaning close to what data mining means-like, mining of knowledge from data, extraction of knowledge, trend (data/pattern) analysis, dredging of data.

Data mining can be used as a synonym to other terms such as KDD (knowledge discovery from data) [2]. Data mining is merely a crucial step in the course of knowledge discovery. The process of discovering knowledge can be depicted in Fig. 1. It involves the series of the following steps, which run iteratively.

(i)   Data cleaning (deals with noise removal and inconsistent data removal)

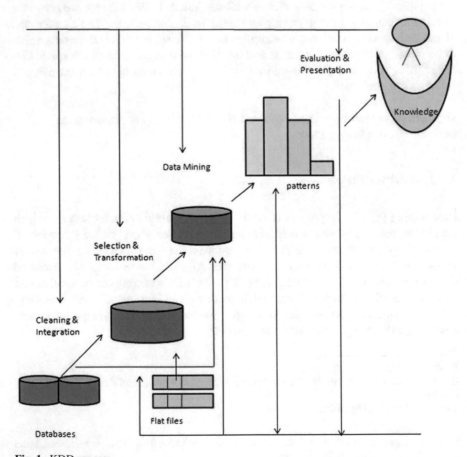

Fig. 1 KDD process

(ii) Data integration (deals with combining of data from multiple data sources)
(iii) Data selection (deals with fetching of data in relevance to the analysis done, fetching of data takes from the database)
(iv) Data transformation (alteration and integration of data into forms appropriate for mining; summary or aggregation operations are performed)
(v)   Data mining (extraction of data patterns using intelligent methods)
(vi) Pattern evaluation (using interestingness measures for the identification of interesting patterns)
(vii) Knowledge presentation (deals with the visualization and knowledge representation techniques for presentation of knowledge mined to users).

Steps (i) through (iv) are concerned with different forms of data preprocessing, where the data is prepared for mining. There may be interaction of the user at the data mining step. The interesting patterns have been identified and presented to the user. Knowledge base is used to store this as a new knowledge.

The former view shows data mining as a step in the KDD (knowledge discovery process). However, in media, in industry, and in the research environment, the term data mining refers to knowledge discovery from data.

Looking at the broader view of data mining functionality "Data mining is the process of identifying interesting knowledge and patterns from vast collections of data". The sources of data can include "databases", "data warehouses", "Web" or other information repositories or the data put into the system in a dynamic manner.

1.1 Types of Data that Can Be Mined

The following is the type of data on which the data mining technique can be applied:

- Heterogeneous and legacy databases
- Data warehouses
- Advanced Database and other information repositories
- Databases comprising of multimedia data
- Databases with text
- Mining text
- Mining WWW
- Object-relational databases
- Transactional databases
- Spatial databases
- Relational databases.

1.1.1 Implementation Process of Data Mining

The process of implementation of data mining [3] (shown in Fig. 2) begins with Business understanding and ends with Deployment.

Fig. 2 Data mining implementation process

(i) Business understanding

This phase deals with setting of business and data-mining goals.

- The first step deals with understanding the objectives of client from business point of view. At this step, requirements are gathered from the client i.e. what the client wants. So, proper formulation of the client's requirement is an essential step.
- There is a requirement to collect of the ongoing scenario of data mining. For this, the assumptions, resources, constraints, and some other relevant parameters are taken care for assessment.
- Data mining goals are defined using business objectives and current scenario.

For the accomplishment of business and data mining goals, the data mining plan should be very detailed.

(ii) Data understanding

In this phase, a check is performed on the data for checking whether it is suitable for the data mining objectives.

- The first step deals with gathering of data using various data sources existing within the organization.
- Flat file, data cubes or multiple databases may be some data sources. Many issues may arise like matching of objects and integration of schema. These issues may come up during the process of Data Integration. The data from various sources are unlikely to match easily, so the process is quite complex and tricky.
- Ensuring that both of the given objects point to the same value or not is really a difficult task. So, Metadata should be adopted for the reduction of errors during the process of data integration.
- Next, properties of acquired data are searched for. For this, some data mining questions are formed as decided in business phase. Query for the same is written. Then to explore properties of data, these questions need to be answered. For answering the questions based on data mining that have been decided in business phase, some tools like query, reporting and visualization tools can be applied. Then depending upon the query results, the quality of data should be confirmed. Also, we need to obtain the missing if present.

(iii) Data preparation

In this phase, data is prepared for use. This process consumes a huge percentage of time i.e. about 90% of the total time of the project. This process deals with collection of data from manifold sources, selection of sources, cleaning of data, transformation, formatting and construction (if required). Data cleaning may be done for data preparation like to clean the data, data smoothing may need to be done and the missing data may be filled.

For example, "age" data is missing for the customers database. Because of the incomplete data, data filling is necessary. Also, data outliers may be there in some cases. For instance, value 200 for the field age. Another issue can be data inconsistency. For instance, different customer name in different tables.

Next step is Data transformation operations that deals with the data and transform it into a meaningful form so that data mining can be applied to it. Several transformations can be applied as discussed in the following text.

(iv) **Data transformation**

Data transformation is a process which would put in towards the accomplishment of the mining process. There are several operations that need to be done for completion of data transformation:

- Smoothing: It is a process of removal of noise from the data.
- Aggregation: It is an operation in which operations like summary or count are applied to the data. Like, on the sales data collected weekly, if aggregation is applied, total is calculated on the monthly and yearly basis.
- Generalization: In this step, concept hierarchies are applied to replace low-level data by higher-level concepts. For example, the city can be generalized to the county.
- Normalization: To perform normalization, the attribute data is scaled up or scaled down. For e.g., data should fall in the range −2.0 to 2.0 post-normalization.
- Attribute construction: The set of attributes helpful for data mining are constructed and included. The output of this process is a data set formed on which modelling techniques can be applied.

(v) **Modelling**

This phase deals with the determination of patterns using mathematical models.

- On the basis of the objectives set up for the business, appropriate modelling techniques need to be chosen for the prepared dataset.
- The quality and validity of the model is checked by creating a scenario.
- Next, the model is made to run on the dataset that has been prepared.
- Then, the results should be assessed by all stakeholders. The major objective is to ensure that the model meets the objectives of data mining.

(vi) **Evaluation**

This phase is concerned with the evaluation of the identified patterns against the business objectives.

- Evaluation of the results which have been generated by adopting the data mining model. The evaluation is done against the business objectives.

- Understanding business requirements is process that runs iteratively. Some business requirements come into picture because of the data mining, so understanding those newly gathered requirements is carried on this step.
- Next step is the deployment phase.

(vii) **Deployment**

- At first, a detailed deployment plan, is created which involves plan for shipping, maintenance, and monitoring.
- Whatever, knowledge or information has been found during the process of data mining must be made understandable for the stakeholders.
- Thereafter, using the key experiences gained during all the phases of a project, a final report is created which works for the improvement of the business policy of the organization.

2 Data Visualization and Its Integration with Data Mining

Visualization is related to the creation of visual images using computer graphics to facilitate understanding of data that is usually massive and complex in nature. Visual Data Mining is the process concerned with discovery of the implicit knowledge from vast sets of data using techniques for visualization. The knowledge discovered is however useful. Visualize can be defined as: "To form a mental vision, image, or picture of (something not visible or present to the sight, or of an abstraction); to make visible to the mind or imagination."

Data Mining is the process of identifying new patterns and insights in data. With increase in the volume of data existing in databases, the need of data summarization, identification of useful but hidden patterns and to act upon the findings arises. Insight gained from data mining can provide terrific economic value which may lead to successful businesses.

The work done in [4] explores issues concerned with the visualization techniques for data mining in business. Especially, it is concerned with the ways for comparison between them. Also, in the work the issues in data visualization are listed and some widely-used visualization techniques are also described. The authors conducted a comparison along with its analysis across visualization techniques.

A number of important visual data mining issues and techniques used to enhance the understanding of the results of data mining has been presented in [5]. In this, the visualization of data before the start of the data mining process and during the process is addressed. Using this process, the quality of the data can be estimated throughout the process of knowledge discovery. It includes data preprocessing, data mining and reporting. Also, information visualization is discussed, which is concerned with how the knowledge uncovered with the help of a data mining tool [28], may be visualized throughout the process of data mining. This aspect includes both, the visualization of the results of data mining as well as the learning process.

There are many aspects and definitions of visualization that are dealt in [6]. The authors have discussed about the visualization process and the steps involved, issues that may arise in visualization, breaking down of visualization techniques based on different perspectives, commonly known data and techniques for information visualization, basic methods for visualization along with their benefits and limitations, process of interactivity and the visualization scope up to some extent in differing research fields.

Data Visualization [7] aids big data for getting an absolute perspective of data as well as for discovering of the values of data. Due to the large number of visualization techniques, it can be difficult to identify the technique that is appropriate for visualization of the data. The aim of visual data representation is to basically give an interpretation to what is insight without difficulties. Different visualization techniques can be applied for carrying out different tasks to communicate different levels of understanding. In this work, various data mining visualization techniques are examined and the adequacy and inadequacy of data visualization technique in handling big data is also discussed.

2.1   Need of Visualization

Data mining algorithms can be used to figure out the hidden patterns in the data. In addition to the mechanical data mining algorithms, visual exploration has proven as an effective tool in data mining and knowledge.

Using data visualization, the data seekers can effectively recognize the Data structural [8] features.

"Data visualization is the process by which textual or numerical data are converted into meaningful images" [9].

As the human brains are very quick in recognizing graphical representations, the data visualization can prove to be an effective tool [10].

Using the visualization techniques to convert the data in raw form into visual graphs, patterns that are hidden in the text and numbers can very easily be detected. The meaning of the patterns can be easily understood by the user more intuitively by the process of effective recognition of the patterns through human brain.

Therefore, the technique of visualization can act as a complement to the data mining techniques.

Data mining and data visualization + huge space in the data warehouse -> source of precious information.

This information gathered can be very important for the business decision makers now days.

2.2 Information and Scientific Visualization

Information visualization and scientific visualization [10] are the two sub-areas combined to form a new discipline termed as Visualization.

Data warehousing and data mining [11] lined the way so that the concept of information visualization can be applied to the high dimensional business datasets.

As an example, the variables in a typical task of scientific visualization are continuous and are about volumes, surfaces, etc.

However, the variables in the task of Information visualization are categorical variables and are concerned with the recognition of hidden but useful patterns, identification of trends, outliers, gaps and cluster identification [12]. The task of data mining in a business data warehouse context has more connection with information visualization.

Data visualization when combined with data mining can prove to be very helpful.

As known that Data mining can be divided into three general phases:

- Phase that deals with data pre-processing,
- Phase applying algorithm on the pre-processed data, and
- Phase that deals with post processing of the result [13].

The pre-processing phase deals with the data visualization. The use of data visualization here can contribute in the selection of attributes and detection of outliers. At the second phase, the visualization techniques help in identification/selection of the algorithm that can be applied to the pre-processed data. The visualization can provide insight and understanding to the result in the third stage that deals post-processing [14].

From the usage point of view, the following are the objectives of visualization tasks:

- Data exploration,
- Hypothesis confirmation, and
- Visual presentation [15].

Depending upon the purpose of task, specific technique of visualization can be picked up based on the requirements.

2.3 Visualization Task Purpose

2.3.1 Data Exploration Task

Geographical layout is one of the most natural means of organizing raw data. The hierarchical feature can easily be observed and understood [16] using this. The technique of treemap and spatial visualization have partial contribution in the exploration task. However, the scatter-plot matrixes are an extension to the 2-D scatter plot diagram for users for observing several 2-D relationships at the same time. For the representation of data from multiple columns at a single screen, survey plots and parallel coordinates are designed [15]. Therefore, the techniques can facilitate more insight into the task of data exploration in business data context.

2.3.2 Hypothesis Confirmation

For confirming a hypothesis already known, the use of traditional 2-D scatter plot seems to be more preferable for the users for recognizing the relationship [12]. Accordingly, the technique of scatter-plot matrices is most helpful in the confirmation of hypothesis scenario because of the 2-D scatter plot presentation format.

Parallel coordinates and survey plots are less useful since the dimensional representation in those two techniques are modified into non-intuitive multiple parallel lines. Treemap and spatial techniques are not necessarily able to represent the hypothesis, thus both are the last choice among the five alternatives.

2.3.3 Visual Presentation

The study of visualization perception is also an important part of information visualization [10]. The objective of visual presentation is to impress and influence the audience. Therefore, the most critical feature in choosing proper visualization techniques is the ease-of perception.

Treemap and spatial visualization techniques can afford the audience with most direct perception to the information carried [16].

The scatter-plot matrices is second to spatial and treemap techniques. The scatter-plot format is easier to grasp then the final batch, i.e. survey plots and parallel coordinates, which require certain amount of explanation to fully recognize the mapping meanings.

In pre-processing phase, the survey plot technique has been proven to perform better than parallel coordinates and scatter-plot matrices in recognizing the important features, exact rules or models [15]. While treemap and spatial visualization are useful only in data with hierarchical and spatial feature, parallel coordinates and scatter-plot matrices can apply to general tasks like outlier detection and attributes selection [15]. With reference to the above, at pre-processing phase of a data mining task, survey plot can be the first choice, followed by scatter-plot matrices and parallel coordinates. Treemap and spatial visualization techniques are only useful under specific scenarios.

In Post-processing phase, after the data mining algorithms performed, the visualization can help check the mined discoveries, or perform new exploration. The pros and cons are similar to the confirmation and exploration tasks mentioned in earlier paragraphs. Thus, the conclusion is identical: parallel coordinates, scatter-plot matrices, and survey plots are better than treemap and spatial visualization.

3 Visualization in Data Mining Applications

The following section discusses how Visualization can be used in Data mining applications.

3.1 Classification with Visualization

Classification: Classification [17, 29] falls in the category of data analysis. It may be defined as a process of identifying a model for the description and differentiation of the data classes and concepts. Say we have a new observation, the Classification task is the problem of identifying the categories, it belongs to. This is done using the training set of data that contain observations along with the categories membership. For example, before starting a project, its feasibility needs to be checked. Two class labels can be used like 'Safe' and 'Risky'. The classifier is required to predict these class labels for adopting the Project and its approval.

Classification involves a two step process:

(i) **Learning Step** (also called Training Phase): This phase is concerned with the construction of model for classification. For building of a classifier, different models are used and the model is made to learn with the help of training set available. For predicting accurate results, the model used has to be trained.

(ii) **Classification Step**: For classification, the model is made for the prediction of the class labels. Then the constructed model is tested on sample data used for testing and then the correctness of the classification rules is estimated.

The work in [18] describes various methods that can be adopted with the classifiers for understanding the structure of the cluster in data.

With the visual tools discussed in [19], one can visualize the class structure in high-dimensional space. This information can further be used to frame a better classifiers which can be a solution to some specific problem. A single data set and classification technique has been used in the work, for real-valued multivariate data, and for gaining insight into other techniques for classification.

3.1.1 Classifiers for Machine Learning

The classifiers for Machine Learning are as follows:

(i)   Decision Trees
(ii)   Bayesian Classifiers
(iii) Neural Networks
(iv)   K-Nearest Neighbour

(v)   Support Vector Machines
(vi)   Logistic Regression
(vii) Linear Regression

3.1.2   Advantages of Classification

Following are the advantages of classification:

- Cost effectiveness of mining based methods
- Efficiency of the mining based methods
- Facilitate identification of criminal suspects
- Facilitates prediction of the risk of certain diseases
- Facilitates banks and financial Institutions in identification of defaulters for the approval of Cards, Loan, etc.

3.1.3   Associated Tools and Languages

These tools facilitated the mining or extraction of the useful information or useful patterns from the raw data.

- Some languages used are. R, Python, SQL etc.
- Major Tools used are: Orange, RapidMiner, Spark, KNIME, Weka.
- Some libraries used are: Pandas, Jupyter, NLTK, NumPy, Matplotlib, ScikitLearn, TensorFlow, Basemap, Seaborn etc.

4   Integrating Visualization with Data Mining in Real Life Applications

- **Market Basket Analysis (MBA)**: This technique is associated with transactions in which some items are purchased in combination. Frequent transactions are considered for the analysis. Example: This technique is used by Amazon and other Retailers. When some product is being viewed by the customers, it may be that the customers are able to view certain suggestions for those commodities as these may have been purchased by some people in the past.
  Representing the market basket analysis graphically facilitates interpreting the whole puzzle of "probabilities/conditional probability/lift above random events" much easier than the tabular format in case of large amount of items.

- **Weather Forecasting**: Weather forecasting is concerned with the changing patterns in weather conditions. This needs to be observed on the basis of some of the parameters such as humidity, temperature, direction of wind etc. For this observation, previous records are required for accurate prediction. It has been presented in [20] that the weather can be forecasted in two ways either using "Short term weather forecasting" or "Long term weather forecasting". Short term forecasting has been implemented for the prediction of the daily forecast. Linear regression is used for Long term forecasting. It is used for predicting the weather trends. The technique is applied on the Real time data. This is implemented in such a way that the data can be presented in a comprehensible graphical format which is user friendly. This resulted in the forecast that in the coming ten years, rain fall will lower down in the northern region of Pakistan. However, an uplift of the humidity has been observed after carefully analysis of the forecasted plots.

- **Detection of Crime related activities by law enforcement agencies**
  Beyond the above applications, Data mining and analytics is used by the crime prevention agencies for identification of the trends across huge data. This is helpful for deciding where the police manpower can basically be deployed. The questions like where is crime most likely to happen and when, whom to search at a border crossing (some factors like category of vehicle used at the time the crime happened, number of people seen at the scene, their age, historical records at the border crossing etc. can be used for this) and which intelligence to be adopted for the counter-terrorism activities.
  A framework has been presented in [21] which transforms the data related to crime into effective visual reports so that the pro-active activities to be done by law enforcement agencies can be strengthened. The input data plays a crucial role in information visualization. In the work, the visualization engine processes whole of the data for the production of the information related to the crime for the law-enforcing agencies. This can be presented in three differing formats: (i) Statistically summarized reports in graphical formats, (2) Heat-maps of crimes and (3) Clusters of crime patterns based on geo-locations.
  Representing the crime related information in a visual manner, may also facilitate the policy-makers for gaining deep knowledge about the types of crime, their timings, the regions of crime etc. The knowledge gained from this, may lead to the improvement of the performance of these agencies for the reduction of the crime rate and for the effective utilization of the resources. Specifically, it performs the processing of crime related information in a comprehensive manner to detect the heat-maps of crimes, clustering of the hidden patterns identified from the criminal data and using information visualization techniques to present it. So, using this system, it is possible to have a comprehensive, chronological and consolidated view

Fig. 3 View of data mining techniques

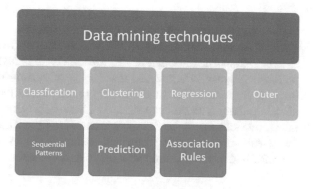

of all forms of criminal activities that have been reported. This is very useful for the law enforcement agencies. Figure 3 shows a view of Data mining techniques.

5 Visualization in Information Retrieval

This field is concerned with the retrieval of information from within a large collection of text documents. Some example systems of information retrieval are discussed below:

- Online Library catalogue system
- Online Document Management Systems
- Web Search Systems etc.

Note: One of the major problems in an IR system, is discovery of the location of the documents within the collection on receiving the query of the user. This type of query of the user comprises of a number of keywords that describe an information need.

In these kinds of search problem, the major focus of the user is to pull out the relevant information from the large document collection. This is suitable for the ad hoc information needs of the user (i.e. short term need). But in case of long term information needs of the user, the retrieval system focuses on the fetching of the fresh information items to the user. Information Filtering is a term used for this type of access to information. The systems corresponding to Information Filtering are referred to as Filtering Systems or Recommender Systems.

Invisible data [22] with their semantic relationships can be transformed into a visible display. Visualization is applicable in the field of information retrieval in two manners:

- To have a visual display of the objects in a meaningful way
- To have a visual display of the process seeking information

As search can be a targeted search and its domain can be complex, lots of cognitive burden may be placed on the user (searcher) to frame and refine their queries, evaluate and explore among the search results, and finally make use of what is found. In such situations, the information visualization techniques may be imprinted to enable searchers to grasp, interpret, and make sense of the information available throughout the search process. The fundamental principles and theories of information visualization along with the explanation of how information visualization can support interactive information retrieval is presented in [23]. This tutorial aims to encourage the researchers for making informed design decisions regarding the integration of the information visualization with the interactive information retrieval tasks.

6 Visualization in Text Mining

The text databases comprises of vast collection of pages. This information is collected from various sources of information like web documents, news articles, e-mails, e-books, digital libraries etc. As this information is vast in amount, the textual databases are spreading at a very quick rate. The data is in semi structured form in the text databases.

For example, there may be few structured fields in documents, like title, author_name, date_publishing etc. Along with these structured components, there are some unstructured components of text also residing in the document. Examples of these components are abstract and contents. For the formulation of effective queries for the analysis and extraction of useful information from the data, the contents of the documents must be known. Further, tools are required to compare the documents on the basis of their rank. Therefore, text mining has gained importance as a famous theme in data mining.

Innovation is the underlying foundation of today's competitive economy and technological advancement. There is a plethora of text mining and visualization tools available on the market for providing the innovative process in discovering "hidden nuggets" of information about the upcoming technologies. A high-level overview of some main tools used for text mining and visualization is presented in this paper [24] for comparing the text mining capabilities, perceived strengths, potential limitations, applicable data sources, and output of results, as applied to chemical, biological and patent information.

Examples of tools discussed incorporate sophisticated text mining software packages, some searching tools working on full-text, and a few data visualization tools that could be combined with the more sophisticated software packages and full-text searching tools.

Web mining [25] is an application of data mining related to the techniques for pattern identification from the WWW. This technique can help in the improvement of the power of web search engine by identification and classification of the web pages. Web mining is a valuable approach that can be applied to commercial websites and e-services.

7 Prediction of User's Next Queries

The paper [26] works for the prediction of the next query that the user may give to the search engine. For this, the proposed system maintains a log of queries that the user has submitted previously to the search engine. These logs are then referred to for gathering this information. The association rule mining [30] technique can then be applied to the queries for the prediction of the query that the user may submit to the search engine in nearby future. Then, the search engine can then keep the responses ready for the predicted queries and the responses can then be provided to the user if her query matches with any of those predicted queries. This increases the efficiency of the search engine.

8 Social Impacts of Data Mining

As almost all the information is available on the web and can be accessed by the interested users and also with the powerful tools of data mining being developed rapidly and put into use, concerns arise that these data mining tools may be a threat to our privacy and security of data. There are many applications of data mining that never play with the user's personal data.

Many studies conducted in data mining research focus to develop some algorithms that are scalable and that do not play with personal data. Data mining focuses on the detection of some significant patterns, not on individual particular information. Privacy issues are concerned with access to individual records without any constraints, like access to credit card transaction records of a customer, medical records of a patient, investigation of criminals etc. If the disclosure controls are not proper or nonexistent, this can be the major grounds for privacy issues. For handling such issues, abundant techniques for enhancing security have been developed. So, this is one of the questions that must be kept at the top of the mind that "What can be done for securing the privacy of individuals during data collection and data mining?".

Multilevel security model can be used in databases for the classification and restriction of data as per the security levels according to which users are allowed to access to only their authorized level. Encryption is a technique in which the data may be encrypted (encoded). One of the techniques that can be used is biometric method for encryption in which the image of a person's iris or thumbprint or fingerprint is used to for encoding of the personal information.

Another active area of research is "Intrusion detection". It deals with not to disclose the sensitive data values and obtain valid data mining results. These data mining methods involve some form of changes to be done on the data for preserving privacy. Such methods lower the granularity of representation for preserving privacy. Like, these methods may work for the generalization of individual data to group. The granularity reduction may lead to the loss of information. There exists a trade-off between loss of information and privacy.

Multimedia data mining is the finding interesting patterns from the multimedia databases. These databases store and manage vast collections of multimedia items. These include image data, audio data, video data and sequence data. These also include hypertext data that contain text, text markups, and linkages.

Multimedia data mining is a field related to several disciplines. Image processing, image understanding, computer vision, pattern recognition and data mining are integrated in multimedia data mining. Similarity search, Content-based retrieval, multidimensional analysis and generalization are also included in multimedia analysis. For storing multimedia related information, additional dimensions and measures are maintained in multimedia data cubes.

9 Graphical Tools

Graphs can be used to display data as it can best describe the values in a shape which is easy to visualize. Also, it can be used to reveal relationships among multiple values in terms of similarities and differences among the data. Graphs are best to describe general trend being followed. These are mainly useful when data is very large.

In addition to this, tabular views of data can also be useful but generally when there is a need to look into normal values. Various types of plots are depicted in Fig. 4.

9.1 Types of Plots

The types of plots are as follows:

- 2D & 3D plots
- Scatter diagram
- Pie diagram
- Histogram
- Bar plot
- Stem and leaf plot
- Box plot

In R, such graphics can be easily created and saved in various formats.

9.1.1 Bar Diagrams

These are used to visualize the relative or absolute values of frequencies of a variable. The bar diagram comprises of one bar for each category.

For each respective category, the absolute frequency or the relative frequency is shown on the *y-axis* and is used to determine the height of each bar.

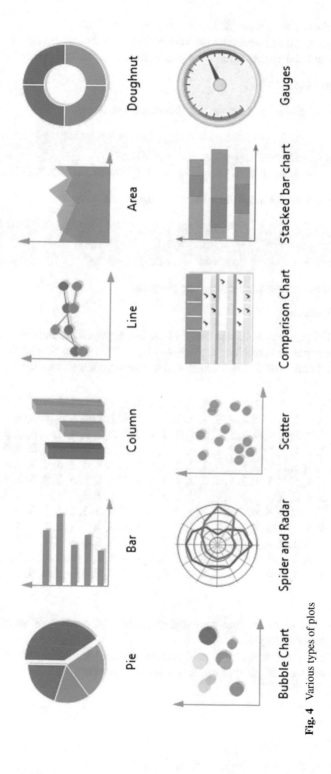

Fig. 4 Various types of plots

However, the width of each bar is arbitrary and not of use.

The following text discusses the plotting of Bar diagrams on R software:

The command Barplot: Creates a bar plot with vertical or horizontal bars.

Detailed command

The command to plot a bar diagram with absolute frequency:

> barplot(table(x)) #Bar plot with absolute frequency

The command to plot a bar diagram with relative frequency:

> barplot(table(x)/length(x)) #Bar plot with relative frequency

Consider the example for plotting bar diagram:

a. Data set *medicine*

A group of 400 patients were randomly given five types of medicines in a clinical trial. The five types of medicines are denoted by 1, 2, 3, 4 and 5. The data on who got which medicine is as follows and stored in a data vector as medicine.

5, 5, 1, 2, 5, 4, 5, 1, 5, 4, 4, 5, 4, 5, 4, 5, 5, 3, 4, 1, 4, 3, 4, 1, 5, 1, 5, 5, 5, 5, 1, 3, 1,
4, 4, 5, 5, 3, 3, 1, 3, 3, 1, 5, 4, 2, 3, 4, 2, 2, 5, 3, 3, 5, 2, 2, 4, 3, 1, 4, 3, 4, 3, 5, 1, 2,
4, 4, 1, 1, 5, 4, 4, 2, 1, 4, 4, 2, 5, 3, 2, 1, 4, 3, 4, 2, 2, 2, 2, 5, 4, 1, 4, 5, 3, 3, 2, 5, 3,
2, 4, 2, 3, 1, 3, 1, 1, 1, 5, 4, 5, 3, 3, 3, 3, 4, 1, 4, 2, 5, 1, 1, 5, 4, 4, 4, 3, 4, 5, 5, 5, 3,
4, 1, 3, 5, 5, 4, 1, 5, 1, 5, 3, 3, 2, 5, 3, 2, 2, 3, 5, 5, 2, 3, 2, 4, 3, 5, 4, 1, 5, 3, 5, 2, 5,
5, 5, 1, 3, 2, 5, 3, 5, 1, 3, 4, 3, 1, 5, 5, 2, 4, 1, 5, 4, 5, 3, 1, 1, 5, 3, 5, 5, 5, 4, 3, 1, 5,
1, 2, 3, 5, 5, 3, 4, 2, 2, 5, 1, 5, 5, 4, 4, 3, 1, 3, 1, 2, 5, 1, 1, 3, 3, 5, 1, 3, 3, 3, 1, 1, 3,
1, 3 , 5, 2, 3, 2, 2, 4, 4, 4, 4, 3, 2, 4, 3, 1, 3, 2, 3, 1, 5, 4, 3, 5, 2, 5, 5, 3, 5, 5, 2, 5, 5,
4, 5, 1, 1, 3, 5, 5, 2, 4, 5, 3, 1, 2, 3, 1, 1, 5, 4, 5, 5, 1, 5, 4, 3, 1, 5, 3, 1, 3, 5, 4, 4, 2,
2, 1, 1, 1, 3, 1, 5, 3, 2, 3, 4, 2, 2, 2, 3, 2, 5, 3, 3, 2, 1, 2, 5, 4, 5, 2, 4, 5, 1, 1, 4, 1, 4,
1, 3, 4, 4, 2, 5, 3, 5, 5, 3, 3, 3, 2, 5, 1, 5, 2, 5, 2, 4, 4, 3, 1, 3, 1, 1, 4, 2, 3, 3, 3, 5, 2,
1, 3, 1, 5, 1, 1, 3, 4, 5, 4, 5, 4, 3, 4, 5, 4, 5, 3, 4, 4, 2, 3, 1, 4, 5, 2, 2, 4, 2, 3, 4, 5, 2,
5, 5, 3, 4

Figures 5 and 6 show the command and screenshot for plotting of bar diagram on the sample data set using R.

Figures 7 and 8 show the command and screenshot for plotting bar diagram on relative frequencies using R on sample dataset.

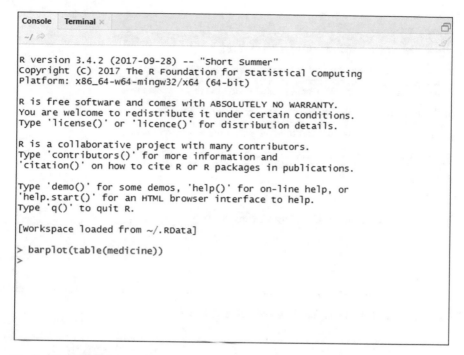

Fig. 5 Plotting bar diagram for the sample dataset

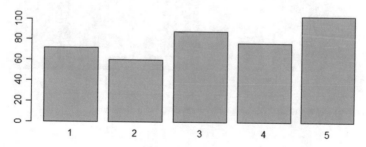

Fig. 6 Screenshot for plotting bar diagram for *medicine* data set on R

9.1.2 Pie Chart

Pie charts visualize the absolute and relative frequencies. A pie chart is represented in form of a circle that is partitioned into segments, where each segments represents a particular category. The size of each segment is dependent upon the relative frequency. The size of each segment is determined by the angle (relative frequency × 360°).

Figures 9 and 10 show the command and screenshot for plotting pie diagram.

```
R version 3.4.2 (2017-09-28) -- "Short Summer"
Copyright (C) 2017 The R Foundation for Statistical Computing
Platform: x86_64-w64-mingw32/x64 (64-bit)

R is free software and comes with ABSOLUTELY NO WARRANTY.
You are welcome to redistribute it under certain conditions.
Type 'license()' or 'licence()' for distribution details.

R is a collaborative project with many contributors.
Type 'contributors()' for more information and
'citation()' on how to cite R or R packages in publications.

Type 'demo()' for some demos, 'help()' for on-line help, or
'help.start()' for an HTML browser interface to help.
Type 'q()' to quit R.

[workspace loaded from ~/.RData]

> barplot(table(medicine) /length(medicine))
> |
```

Fig. 7 Command for plotting bar diagram on relative frequencies

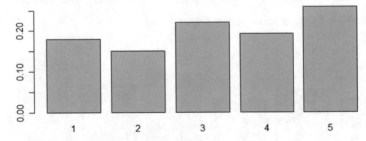

Fig. 8 Screenshot for plotting bar diagram on relative frequencies

9.1.3   Histogram

In Histogram, the data is categorized into different groups. In this, a bar is plotted for each category may be with differing height. Data is continuous. The area of the bars (=height × width) is proportional to the frequency (or relative frequency). So the widths of the bars may change in accordance with the frequency.

Figures 11 and 12 show command and screenshot for plotting Histogram.

9.1.4   Kernel Density Plots

In histogram, the continuous data is artificially categorized. Choice and width of class interval is crucial in the construction of histogram. Other option is kernel

```
R version 3.4.2 (2017-09-28) -- "Short Summer"
Copyright (C) 2017 The R Foundation for Statistical Computing
Platform: x86_64-w64-mingw32/x64 (64-bit)

R is free software and comes with ABSOLUTELY NO WARRANTY.
You are welcome to redistribute it under certain conditions.
Type 'license()' or 'licence()' for distribution details.

R is a collaborative project with many contributors.
Type 'contributors()' for more information and
'citation()' on how to cite R or R packages in publications.

Type 'demo()' for some demos, 'help()' for on-line help, or
'help.start()' for an HTML browser interface to help.
Type 'q()' to quit R.

[workspace loaded from ~/.RData]

> barplot(table(medicine) /length(medicine))
WARNING: You are configured to use the CRAN mirror at https://cran.rstudio.com
/. This mirror supports secure (HTTPS) downloads however your system is unable
 to communicate securely with the server (possibly due to out of date certific
ate files on your system). Falling back to using insecure URL for this mirror.

To learn more and/or disable this warning message see the "Use secure download
 method for HTTP" option in Tools -> Global Options -> Packages.
> pie(table(medicine))
```

Fig. 9 Command to plot pie diagram

Fig. 10 Screenshot for plotting pie chart for *medicine* data set on R

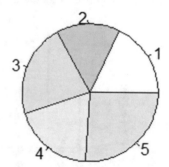

density plot. It is a smooth curve and represents data distribution. Density plots are like smoothened histograms. The smoothness is controlled by a parameter called bandwidth. Density plot visualises the distribution of data over a continuous interval or time period. Density plot is a variation of a histogram that uses kernel smoothing to smoothen the plots by smoothing out the noise.

Figures 13 and 14 show command and screenshot for plotting kernel density plot for sample data set.

```
R version 3.4.2 (2017-09-28) -- "Short Summer"
Copyright (C) 2017 The R Foundation for Statistical Computing
Platform: x86_64-w64-mingw32/x64 (64-bit)

R is free software and comes with ABSOLUTELY NO WARRANTY.
You are welcome to redistribute it under certain conditions.
Type 'license()' or 'licence()' for distribution details.

R is a collaborative project with many contributors.
Type 'contributors()' for more information and
'citation()' on how to cite R or R packages in publications.

Type 'demo()' for some demos, 'help()' for on-line help, or
'help.start()' for an HTML browser interface to help.
Type 'q()' to quit R.

[Workspace loaded from ~/.RData]

> barplot(table(medicine) /length(medicine))
WARNING: You are configured to use the CRAN mirror at https://cran.rstudio.com
/. This mirror supports secure (HTTPS) downloads however your system is unable
 to communicate securely with the server (possibly due to out of date certific
ate files on your system). Falling back to using insecure URL for this mirror.

To learn more and/or disable this warning message see the "Use secure download
 method for HTTP" option in Tools -> Global Options -> Packages.
> pie(table(medicine))
>
> hist(marks,main = "Marks of Students")
```

Fig. 11 Command for plotting histogram

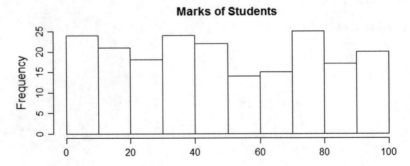

Fig. 12 Screenshot for plotting histogram for *marks* dataset on R

```
>
>
>
>
> plot(density(marks, kernel='gaussian'), main="Density of Marks")
>
> |
```

Fig. 13 Command for plotting Kernel density plot

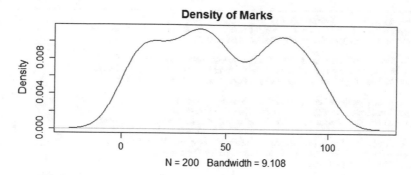

Fig. 14 Screenshot for plotting kernel density plots for *marks* dataset

9.1.5 Boxplot

Box plot is a graph which summarizes the distribution of a variable by using its median, quartiles, minimum and maximum values. It can be used in comparing different datasets.

boxplot() draws a box plot.

Following dataset yield (in kilograms) are reported from 200 agricultural fields of same size. The data is stored in a data vector yield.

34.4, 47.0, 19.6, 20.9, 39.4, 40.1, 47.2, 28.5, 44.4, 22.5, 18.3, 46.8, 12.1, 26.4, 28.3, 26.6, 36.8, 40.3, 46.5, 42.8, 13.7, 17.1, 35.7, 45.0, 33.7, 20.5, 45.4, 17.5, 29.6, 10.4, 24.4, 27.7, 15.0, 35.0, 22.1, 19.6, 24.3, 45.7, 49.6, 39.3, 49.7, 31.6, 27.4, 48.6, 15.9, 12.7, 11.0, 34.5, 37.9, 42.0, 15.5, 16.4, 49.6, 25.9, 17.5, 29.1, 31.8, 23.1, 50.0, 31.1, 15.3, 27.5, 34.8, 18.1, 15.4, 41.1, 35.4, 21.3, 17.7, 20.6, 31.2, 37.4, 25.3, 48.2, 14.7, 11.6, 30.2, 33.1, 43.6, 36.2, 47.8, 30.5, 13.4, 49.8, 26.1, 45.8, 45.1, 21.9, 15.3, 20.6, 10.2, 42.8, 17.0, 43.7, 16.7, 40.6, 30.8, 20.9, 23.7, 38.2, 33.7, 28.8, 23.5, 48.7, 35.8, 17.9, 24.3, 30.5, 45.3, 16.1, 19.2, 16.5, 34.6, 30.1, 17.5, 26.3, 33.3, 22.4, 29.2, 47.6, 11.8, 31.4, 27.7, 46.3, 45.2, 16.5, 40.1, 26.1, 32.3, 13.2, 14.7, 47.0, 45.2, 16.5, 31.3, 47.2, 23.0, 16.4, 48.0, 28.5, 18.8, 10.1, 34.8, 26.1, 46.0, 30.2, 39.1, 11.1, 25.2, 25.5, 23.5, 24.6, 35.6, 11.3, 37.8, 42.6, 30.3, 14.5, 46.3, 26.5, 29.0, 38.5, 19.7, 22.0, 38.2, 40.9, 10.6, 32.1, 36.1, 47.3, 37.6, 20.2, 26.4, 14.9, 15.3, 35.6, 23.9, 26.9, 47.6, 25.4, 19.1, 37.6, 10.4, 37.4, 41.7, 30.3, 22.3, 39.5, 22.2, 41.0, 14.5, 41.9, 29.6, 43.3, 40.3, 46.1, 21.1, 27.8, 20.9, 23.2

Figures 15 and 16 show the command and screenshot for plotting Boxplot on sample data set.

```
R version 3.4.2 (2017-09-28) -- "Short Summer"
Copyright (C) 2017 The R Foundation for Statistical Computing
Platform: x86_64-w64-mingw32/x64 (64-bit)

R is free software and comes with ABSOLUTELY NO WARRANTY.
You are welcome to redistribute it under certain conditions.
Type 'license()' or 'licence()' for distribution details.

R is a collaborative project with many contributors.
Type 'contributors()' for more information and
'citation()' on how to cite R or R packages in publications.

Type 'demo()' for some demos, 'help()' for on-line help, or
'help.start()' for an HTML browser interface to help.
Type 'q()' to quit R.

[Workspace loaded from ~/.RData]

WARNING: You are configured to use the CRAN mirror at https://cran.rstudio.com
/. This mirror supports secure (HTTPS) downloads however your system is unable
 to communicate securely with the server (possibly due to out of date certific
ate files on your system). Falling back to using insecure URL for this mirror.

To learn more and/or disable this warning message see the "Use secure download
 method for HTTP" option in Tools -> Global Options -> Packages.
> boxplot(yield)
>
```

Fig. 15 Command for plotting Boxplot on sample data set

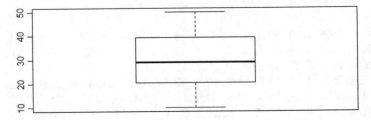

Fig. 16 Screenshot for plotting Boxplot for *yield* dataset on R

10   Conclusion

For drawing a particular conclusion out of the existing complex data, there is no need of studying any theoretical results. Data mining and data visualization can be integrated with each other to have advantages in the area of data science applications in computer science field. The chapter focuses on data mining, the KDD process, types of data that can be mined and its implementation process. The complete visualization process along with its integration with data mining is presented. The benefits of visualization techniques in data mining applications in real life are listed. The chapter also discusses the usage of visualization in domains of information retrieval and text mining. The chapter details the graphic tools that can be used to visualize the data. Each graphical tool is elaborated. The command that can be used for plotting the graphs using R software are listed with snapshot to show how the results can be obtained as the result of executing the command.

References

1. Maurizio M (2011) Data mining concepts and techniques. E-commerce. (http://www.dsi.unive.it/~marek/files/06%20-%20datamining)
2. https://www.tutorialspoint.com/data_mining/dm_knowledge_discovery.htm
3. https://www.guru99.com/data-mining-tutorial.html
4. Yeh RK (2006) Visualization techniques for data mining in business context: a comparative analysis. In: Proceedings from decision science institute, pp 310–320
5. Viktor HL, Paquet E (2008) Visualization techniques for data mining. In: Data warehousing and mining: concepts, methodologies, tools, and applications. IGI Global, pp 1623–1630
6. Khan M, Khan SS (2011) Data and information visualization methods, and interactive mechanisms: a survey. Int J Comput Appl 34(1):1–14
7. Ajibade SS, Adediran A (2016) An overview of big data visualization techniques in data mining. Int J Comput Sci Inf Technol Res 4(3):105–113
8. Nabney IT, Sun Y, Tino P, Kaban A (2005) Semi-supervised learning of hierarchical latent trait models for data visualization. IEEE Trans Knowl Data Eng 17(3):384–400
9. Post FH, Nielson GM, Bonneau G (eds) (2003) Data visualization: the state of the art. Kluwer Academic Publishers
10. Ware C (2004) Information visualization: perception for design, 2nd edn. Morgan Kaufmann, San Francisco
11. Han J, Kamber M (2000) Data mining: concepts and techniques. Morgan Kaufmann
12. Shneiderman B (2003) Why not make interfaces better than 3D reality? IEEE Trans Comput Graph Appl 23:12–15
13. Bernstein A, Provost F, Hill S (1998) Toward intelligent assistance for a data mining process: an ontology-based approach for cost-sensitive classification. IEEE Trans Knowl Data Eng 17(4):503–518 (Bishop CM, Tipping ME)
14. Witten IH, Frank E (2016) Data mining: practical machine learning tools and techniques with java implementations. Morgan Kauffmann
15. Grinstein G, Ward M (2002) Introduction to data visualization. In: Fayyad U, Grinstein G, Wierse A (eds) Information visualization in data mining and knowledge discovery. Morgan Kaufmann, California
16. Soukup T, Davidson I (2002) Visual data mining: techniques and tools for data visualization and mining. Wiley
17. Bhowmick SS, Madria SK, Ng WK (2003) Web data management. Hardcover, Springer
18. Cook D, Caragea D, Honavar V (2004) Visualization for classification problems, with examples using support vector machines
19. Cook D, Caragea D, Honavar H (2004) Visualization in classification problems. In: COMPSTAT' symposium
20. Bibi N (2017) Web based weather visualization and forecasting system. Gazi Univ J Sci 30(4):152–161
21. Shah S, Khalique V, Saddar S, Mahoto NA (2017) A framework for visual representation of crime information. Indian J Sci Technol 10(40). https://doi.org/10.17485/ijst/2017/v10i40/120079
22. Zhang J (2008) Visualization for information retrieval. https://doi.org/10.1007/978-3-540-75148-9
23. Hoeber O (2018) Information visualization for interactive information retrieval. In: Proceedings of the 2018 conference on human information interaction & retrieval. ACM, pp 371–374

24. Yang Y, Akers L, Klose T, Yang CB (2018) Text mining and visualization tools–impressions of emerging capabilities. World Patent Inf 30(4):280–293
25. https://www.tutorialride.com/data-mining/web-mining.htm
26. Madaan R, Sharma AK, Dixit A (2015) A data mining approach to predict users' next question in QA system. In: 2015 2nd international conference on computing for sustainable global development (INDIACom). IEEE, pp 211–215
27. Pujari A (2010) Data mining techniques, 2nd edn. Orient BlackSwan/Universities Press
28. https://opensourceforu.com/2017/03/top-10-open-source-data-mining-tools/
29. https://www.tutorialspoint.com/data_mining/dm_classification_prediction.htm
30. https://www.cs.upc.edu/~mmartin/D7-%20Association%20rules.pdf

Relevant Subsection Retrieval for Law Domain Question Answer System

Aayushi Verma, Jorge Morato, Arti Jain and Anuja Arora

Abstract Intelligent and instinctive legal document subsection information retrieval system is much needed for the appropriate jurisprudential system. To satisfy the law stakeholders' need, the system should be able to deal with the semantics of law domain content. In this chapter, a sophisticated legal Question-Answer (QA) system is developed specifically for law domain which will be able to retrieve the relevant and best suitable document for any specific law domain queries posted by users'. Legal QA system is developed with the help of two computational areas—Natural Language Processing and Information Retrieval. This system is developed in an amenable way to retrieve the relevant subsection in accordance with the legal terminology embedded inquiry entered by the user. Syntactic and semantic analysis of legal documents followed by query processing helps in retrieving inferences from the knowledge base to answer the query. In our research, various models have been analyzed in the opinion of the document matching threshold value. Satisfactory results are obtained by 0.5 threshold value.

Keywords Information retrieval · Natural language processing · Word2vec · TF-IDF · Part-of-speech tagging · Legal system · Law domain

A. Verma
Hays Business Solutions, Noida, India
e-mail: aayushi291091@gmail.com

J. Morato
CSE, University Carlos III de Madrid, Leganes, Spain
e-mail: jorgeluis.morato@uc3m.es

A. Jain (✉) · A. Arora
CSE, Jaypee Institute of Information Technology, Noida, India
e-mail: arti.jain@jiit.ac.in

A. Arora
e-mail: anuja.arora@jiit.ac.in

© Springer Nature Switzerland AG 2020
J. Hemanth et al. (eds.), *Data Visualization and Knowledge Engineering*,
Lecture Notes on Data Engineering and Communications Technologies 32,
https://doi.org/10.1007/978-3-030-25797-2_13

1   Introduction

Today's age is known as Information age which is having no fear of running out of knowledge/information in any domain. Knowledge and Information resources are increasing exponentially in all the directions and domains. Therefore, exponential information growth has become a challenge and overwhelmed the domain experts. Domain experts want to get the useful, relevant and entire content according to their desire. Lots of online law domain knowledge sources are available in web community and researches may make use of these resources to handle various information management related problems of considered domain. According to Mitratech,[1] an enterprise legal software provider has done a survey and released that law departments budget will expand by 18% for legal Knowledge Management (KM) systems in coming years. Legal knowledge management is basically on rising stage and law firms started knowledge strategy to support their system. As an innovative application [1, 2] legal document retrieval system has been chosen in this research work. Certainly, this proposed technological innovation of law domain knowledge discovery and law information resource management helps law stakeholders to retrieve relevant documents according to their need. Legal document retrieval system is an important and commanded information resource management system which is today's need. This is needful for users as well as legal persons those search for a specific problem related document from the collection of law knowledge source.

Legal knowledge management helps law firms and legal domain clients by providing efficient and effective operations. Legal KM is an essential approach to provide a solution while maintaining quality. Even, the legal system has also become part of this information challenge. On one end, majority of legal services relevant to internet contains legal documents such as 'terms and condition' section, 'copyright agreement', 'notices' etc. On the other end, legal system must deal with the relevant document retrieval problem while preparing for legal cases in the daily scenario. Law firms require a system which can map the law sections' content to answer the user query. In the current scenario, law firms need to dig into number of documents to find out the most pertaining document in support of specific law case/law inquiry.

Most recent and relevant work in this direction is done by Boella et al. [3] using the Eunomos software which is an advanced legal document and knowledge management system. Their system is used to answer all legal queries based on legislative XML and ontology. In this chapter, the proposed system is also developed on the same fundamentals and tries to cover legal research challenges such as complex, multi-level, multi-lingual, and information overload. In 2017, Cowley [4] have published in their work—the experiments that are performed to provide expert legal advice to an individual as well as business firms by sharing legal domain information while using an effective knowledge management to provide the solution to UK law firms.

Computer science usage to inform legal decision making has become one of the most desired and demanded goal for the field of law. Law domain information is not

---

[1]http://www.mitratech.com/.

well structured from Knowledge Engineering (KE) perspective. Consequently, both data construction and information retrieval are not straight forward processes. Inferring concept and knowledge from online available law domain data is the complex process for decision support system perspective [5, 6]. The legal knowledge management systems are composed of legal terms collection and relationship among them. The legal management system answers users' query according to associate concepts and articles of legislation process.

Basically, the content size, variety and quality of the knowledge source provided to a decision support system proportionally affect the impact of quality of outcome. Surely, to improve the legal decision making scenario with the help of computational system is a daunting task. Managing and Monitoring legal content from the ocean of ever increasing legal document (legal contracts, agreements, legal affair, claim deed, etc.) is a cumbersome task. Henceforth, retrieval of most relevant/matching legal document according to specific query is still under progress by researchers. Usually other domain specific systems are digitized on fast pace, but retrieval of legal document is a manual and labor-intensive task even in recent time. People are performing this activity manually which undoubtedly will not provide entire as well as accurate content i.e. might fetch incomplete information using manual process. Furthermore, law domain users must adhere an effective plan to retrieve specific content from their enormous manually kept data achieves. Even, problems exist in the smooth and precise functioning of the system to retrieve relevant content from digital law domain achieves.

According to the studies, segmenting topic related questions into parts can help the retrieval system to fetch the most appropriate answer that are most relevant to the user's query [7]. Even Part-of-Speech (POS) tagging has been used to improve result quality. POS tagging is a process of tagging up the word corresponding to the particular part of speech [8] which helps to identify and explore the relationship between adjacent words as nouns, verbs, adjectives, adverbs etc. The key idea of the user's query can be mainly relied on nouns and verbs and hence using POS tagging gives more meaning to the sentence and reflects little improvement in precision.

This article focuses on legal Question-Answer (QA) system in context of legal knowledge management. Knowledge management is considered broadly, to include the consequence of data available for law domain. A well-formed law domain question answer system is requisite to solve this problem. Retrieval outcome in form of inappropriate and incorrect document of information carries the risk of underpinning knowledge. There has been growing research interest in building question answer system, these kinds of systems are either domain restricted or open domain. Importance of law domain system is increased due to increasing amount of law domain content is now available through electronics medium. Legal system performance may degrade because of the following document related issues.

- Scalability: Legal domain is kept on increasing its content scope, size and complexity;
- Different source of legal norms such as regional, national, international etc.;

- Redundant Information: Some enacted laws are modified or overridden according to recent norms;
- Boundaries: Many laws are applicable on varying subjects, so retrieve document within the scope/boundaries.

This chapter projects an approach which is an effort to overcome legal document accessibility and searchability issue as it is not kept at pace in this growing information era. Accurate legal QA system is important to provide access to laymen and legal professionals. Various Natural Language Processing (NLP) and Information Retrieval (IR) techniques have been used to improve performance of application and have been built to fulfill law stakeholders' requirement. The chapter focuses on automation of various activities involved within the law domain services. Various techniques are placed in a stacking manner which gives birth to various multilevel models. Further legal system performance for all level models is analyzed and recorded.

Above mentioned are few legal document related issues. Whereas, there exist some technical retrieval related issues also. Thus, we have formed some research questions which are answered in this chapter to make a further step in effective legal document retrieval system under mentioned document and retrieval issue constraints.

To address the above issues, a relevant subsection retrieval model and corresponding approach is developed which is fundamentally a combination of information retrieval and natural language processing techniques. The easiest method in the QA system is to respond to a specific user input according to the available response list. As we do not have and direct response list, therefore retrieval system has been developed to provide relevant response to user according to their need using our proposed approach. The QA system which is implemented in this research work is a composite approach and included Word2Vec and Term Frequency (TF)—Inverse Document Frequency (IDF) approach to provide advancement to QA model.

The rest of the chapter is organized as Sect. 2 presents the literature studied due to understand and solve the relevant subsection retrieval problem. Some state-of-art approaches for legal information retrieval system are also detailed in this section. Section 3 introduces the dataset which is used to perform experiment and validate approach. Section 4 provides overall work done to achieve high quality results. Section 5 depicts the empirical evaluation i.e. results are validated and presented. Finally, Sect. 6 concludes the chapter and discusses future directions.

2 Related Work

There is an immense wealth of relevant information retrieval literatures in the form of research papers and articles that suggest numerous distinct expositions for a competent, effective and feasible solution. Various researchers have contributed in this direction and published their work. Based on the literatures, the current state of art for information retrieval in legal domain has been put forth into three major classes—Knowledge Engineering based Retrieval System, Natural Language based Retrieval System, and Information Retrieval System. These are discussed in detail in this section.

Direct use of computerized information retrieval faces multifaceted problems - heterogeneity in the domain relevant types, structures, languages, and terminologies of documents, amongst others. Lyytikäinen et al. [9] have showcased problems w.r.to European Legal system's information retrieval framework when are encompassed within the computational architecture. They have used EULEGIS project for this task.

Rather than a general search engine as used to search the web, domain specific information retrieval is gaining importance due to their specific structure, keywords, and context [10]. This leads to dimension-based information retrieval system, an interface for retrieval of relevant information pertaining to human anatomy. Such usage is shown to avoid complex hierarchical framework.

Legal document retrieval, another domain-specific IR system, is thriving as a budding research topic with various other retrieval domains. This situation has probably contributed to the fact that legal retrieval systems have wider perspectives and so deserves more attention. Legal documents mostly contain text pertaining to statutes, regulations, cases, precedents, legal literature, contracts, and other specific literatures. In this domain, the necessary information to be retrieved is usually scattered in different documents, sometimes in different data banks, and mapping the links among the required pieces of information may be difficult to establish [11].

The initial impetus of question classification came into focus in well-known Text RetriEval Conference (TREC) as a specific QA track [12]. In [13] authors have proposed a machine learning approach for classification of questions and their requirements. A two-layered taxonomy approach is used—coarse level and classification level while using TREC's natural semantic classification.

Law domain is considered as one of the original domains to embrace information retrieval into its folds as it is a text-based discipline that follows requirement of multiple case study-based arguments to support a clause or court-case under consideration for argument or judgment process.

Based on existing legal domain specific literature, approaches are classified in three categories—(1) Knowledge engineering based approaches [14, 5, 15, 16, 17, 12]; (2) Natural language processing based approaches [18, 11, 19, 20, 21, 22]; and (3) Information retrieval based approaches [3, 23, 24, 25, 26, 27].

2.1 Knowledge Engineering Based Retrieval System

Manual knowledge engineering frameworks and ontologies are required for reflecting proper recall of case-studies and leveraging legal precedents. This attempts to translate to retrieval system from manual method of remembering legal aspects to its content and significance. Considering highly immersive environments with interaction through multiple modalities, the tracking of such knowledge becomes even more complex. Such environments have been increasingly used to support decision-making practices, which may involve cognitive-intense activities and critical thinking. Inferring concepts and knowledge from logging data in such activities is key for improving design of decision support systems, and general systems as well. Knowledge engineering based approaches for legal retrieval systems are limited to artificial intelligence and case-based reasoning approaches [12]. Whilst, knowledge ontology and knowledge graph based approaches are increasingly used to support other domain retrieval systems. Conceptual query expansion was proposed by authors in [5]. They addressed issues and open methodological questions relevant to core legal ontology structure. Search query generation using a conceptual semantic network and synonym set are dealt in [15]. Marking the named entities [28, 29], ranking of the retrieved documents and selection of final set of documents and forming conceptual semantic network for giving appropriate and useful results to the query are detailed. Kanapala et al. [16] have shown approaches for the tasks of indexing and retrieval which have used language modelling and inference networks.

In the architecture as is proposed by Quaresma and Rodrigues [20], the authors have used partial knowledge retrieval system for information extraction and semantic and pragmatic interpretation.

2.2   Natural Language Based Retrieval System

NLP based retrieval systems were initially used in conjunction with case-based retrievals. Inference networks augment the KE system by incorporating additional weighted indexing [22]. This allowed the framework not only to support multiple document representation schemes and amalgamation of results from multiple and different queries and query types to be combined, but also facilitated flexible match between the terms and concepts mentioned in query tokens.

In [11], Pietrosanti and Graziadio proposed a framework that represents the context as structural and functional dimensions. This would ensemble the hierarchical organization of legal documents and the web is given by the legal cross-references as well as provide content descriptors. Their work encompasses (i) Acquisition of information from legal documents, (ii) Intelligent search and navigation through legal databases, and (iii) Legal document drafting support. The model describes the contextual and functional requirements based on the document's concept-anchor structure, hierarchy and classification of definition. The juridical style of text contains typographical layout, formal and recurrent expressions and a specialized vocabulary constrained by legal jargons, hence they have much specific morphological, syntactic and semantic textual forms which help in forming the functional rule sets. Various techniques of NLP such as n-grams, bag of words, and TF-IDF are explored in details. Since the retrieval is from a task-specific corpus, n-gram model gave better results. The authors have used word clustering [30] and textual entailment by incorporating machine learning techniques such as string matching, word ordering, word overlap, POS overlap, dependency parsing and deep learning.

In [5] also, the authors have depicted semantic information in legal ontology systems and their usage for question-answer applications. NLP methods have known to be useful in certain areas and pre-processing of queries in such cases. Multilingual experimentation are also conducted in [26] without getting much valued success. Such question-answer system is also used in [20] wherein syntactical analysis stage uses NLP methods. To fine-tune the system, ontology and knowledge representations are used in conjunction to logic programming, and pragmatic interpretation are performed by considering the referred ontology parameters. Answer generation from queries utilized referent variables of the question, their conditions and knowledge base.

As shown by Peruginelli [19], multi-lingual legal information access is in much demand due to its requirement in language integrity across globe and cultural diversity maintenance. Two such systems—MultiLingual Information Retrieval (MLIR) and Cross-Language Information Retrieval (CLIR) are in much demand to help dismantle large scale digital collections across geography and multiple cultures. Legal translations' significance is manifold due to its interactions and influence in all spheres of life and governance. The functional equivalence of legal jargons and their semantic and syntactic relevance are of grave importance.

In the construction of appropriate semantic ontology, use of NLP approach like partial parsers and lexical information are discussed in [21]. They have shown the process to consist of structural objects and semantic objects and use inference engine to conclude their query. Because most systems of legal querying resemble question-answer sessions, advanced semantic techniques need to be pursued [31]. A cognitive model of legal document analytics is presented in [18] which have experimented with service agreement document sets.

2.3 Information Retrieval System

The most commonly used methods for generating responses is retrieval based models which primarily uses a repository of predefined responses and heuristic to pick an appropriate response based on the input and context. The heuristic could be as simple as a rule-based expression match, or as complex as an ensemble of machine learning classifiers. These systems don't generate any new text; they just pick a response from a fixed set.

Adebayo et al. [14] used similarity matching method, namely—word embedding similarity [28], WordNet similarity, negation words, and vector space based similarity [29]. Opijnen and Santos [27] have highlighted how the information retrieval systems are steeply used in law such as Legal Expert System (LES) and Legal Information Retrieval (LIR). Both these systems have high relevancy and are used extensively in the interaction between the end-user, legal domain and IRS. These relevancies are visible in the seven main aspects, namely—algorithmic relevance, bibliographic relevance, topical relevance, cognitive relevance, situational relevance and domain relevance. The holistic requirement of relevance to retrieve solution for a query in the legal domain has been explored.

Moulinier and Molina-Salgado of Thomson Legal and Regulatory Research and Development Group have depicted their experience with dictionaries and similarity thesauri for the bilingual task, which are used for machine translations in their multi-lingual runs for query [26]. Translation between languages may result in changes of legal jargons, their context and requirements. The WIN system [23] is evolved from an inquiry system which is extensively used in monolingual, bilingual and multi-lingual experiments. Certain result parameters give more insight into its usefulness. Timely retrieval of relevant and useful documents is important for survival in legal domain [25]. TREC legal data set are used to retrieve document requirements. Apart from document normalization and analysis, query pruning also need to be articulated.

Law definition, classification, jurisdiction, volume, accessibility, updates and con-solidated text, its context, understanding, usage and interpretation such are the chal-lenges in retrieving relevant documents and examples in legal domain. Framework for legislative XML and legal ontologies and informatics require not only storing and retrieving legislation but also traversing legal terminology, representing norms, reasoning and argumentation using Eunomos [3] system. Kim et al. [24] have pro-posed a methodology for searching statutes in conjunction with the relevance scores to determine the production of R&D projects. Their searching process have used a morpheme analyzer to select only nouns and its sets of statutes definition are followed by the network connectivity based ranking.

In this work, several prevalent models are analyzed and the appropriate functional context information in the framework are used to extract improved retrieval accuracy.

3 Law Domain Datasets

To experiment, British Columbia Security Commission (BCSE) is been taken which is basically a regulatory agency content which administers and enforces legislation in British Columbia. Therefore, BCSE data is crawled from BCSE website[2] and all the seven sections are extracted as listed in Table 1.

This data is basically an unstructured data which need to be changed in a structured form. Data descriptive report is detailed in Table 1 which shows individual word count in the individual sections. Snapshot of how data looks like is shown in Fig. 1 which depicts 'Security Acts' content. Overall taken dataset contains 356 pages.

---

[2]https://www.bcsc.bc.ca/Securities_Law/.

Table 1 Dataset statistics

S. no.	Section	Number of words
1	Securities act	150,000
2	Securities regulations	30,000
3	Securities rules	50,000
4	National instruments	90,000
5	Securities policies	50,000
6	Forms	75,000
7	Notices	30,000

1 (1) In this Act:

"**adviser**" means a person engaging in, or holding himself, herself or itself out as engaging in, the business of advising another with respect to investment in or the purchase or sale of securities or exchange contracts;

"**associate**" means, if used to indicate a relationship with any person,

(a) a partner, other than a limited partner, of that person,

(b) a trust or estate in which that person has a substantial beneficial interest or for which that person serves as trustee or in a similar capacity,

(c) an issuer in respect of which that person beneficially owns or controls, directly or indirectly, voting securities carrying more than 10% of the voting rights attached to all outstanding voting securities of the issuer, or

(d) a relative, including the spouse, of that person or a relative of that person's spouse, if the relative has the same home as that person;

"**auditor oversight body**" means a self regulatory body that

(a) regulates the auditing or review of financial statements that are required to be filed under this Act, and

(b) is recognized under section 24;

"**business day**" means a day other than Saturday or a holiday;

"**Business Development Bank of Canada**" means the Business Development Bank of Canada incorporated under the *Business Development Bank of Canada Act* (Canada);

"**class of exchange contracts**" includes a series of a class of exchange contracts;

"**class of securities**" includes a series of a class of securities;

"**clearing agency**" means a person who

(a) in connection with trades in securities, acts as an intermediary in paying funds, in delivering securities or in doing both of those things,

(b) provides centralized facilities through which trades in securities or exchange contracts are cleared, or

(c) provides centralized facilities as a depository of securities;

Fig. 1 Security Act dataset snapshot

4 Proposed Methodology

With the advent of internet, mobile platforms and numerous services, the footprints of domain content are easily available in the online world for every domain. Even, content related to targeted domain in this chapter is also available in high volume. Therefore, experiments to build a Question Answer system for law domain are performed to cover two perspectives

- Legislative perspective i.e. for law domain stakeholder such as lawyer, judge;
- General perspective i.e. for raw user/client/end user.

In particular, if stakeholder asks any question relevant to law, our legal system is capable enough to retrieve the subsections from law knowledge base along with the order of relevance. And if, system gives information to a raw user/client/end user then the system provides simplified outcome to the user primarily using the web content/web responses as more influential than legal informative summarization. So, varying stakeholders can use this system with varied users' perspective. System is developed for varying usage perspective such as for lawyers the system accumulate information from various legal documents and present it with an inferential reasoning. On the other end, for end users the Google corpus is used.

This section comprises of two Sects. 4.1 and 4.2. Section 4.1 discusses overall model that is used to retrieve the law domain. Section 4.2 details an approach where all the techniques that are used to achieve objective are elaborated.

4.1 Relevant Subsection Retrieval Model

With respect to the legislative perspective, the retrieval uses law domain corpus to generate useful results. After preprocessing of the query, the process follows a four-stage model:

- Building recursive tree using the query tokens;
- Storing relevant query requirements in the data frames;
- Utilization of legislative corpus Word2Vec Model; and
- Tokenization for relevant text retrieval.

The Word2Vec model follows either the contiguous bag-of-words or contiguous-skip-gram architectures. The context is for the legal domain document retrieval, thus the model uses a hybrid solution which blends contiguous-skip-gram approach as it relies on the usage of near words to understand the context. The model starts with the preprocessing such as token generation, followed by POS tagging, generation of sentence vectors and similarity measure of the vectors. Preprocessing further involves removal of stop words such as articles, prepositions etc., lemmatization, stemming and utilization of frequency filters. Thus reduce the query criterion to a manageable set of parameters—suitable training method, sub-sampling of high-frequency words, quality of words by embedding dimensionality, size of context words, amongst others. The method generates an overall similarity score that is used to match appropriate keywords from the legal domain corpus, hence resulting in recoupment of relevant and useful documents containing case-studies etc. Figure 2 shows overall relevant subsection retrieval model which is divided into two boundaries—one boundary shows relevant subsection retrieval from the law corpus, whereas, another boundary shows from the Google corpus—general user perspective.

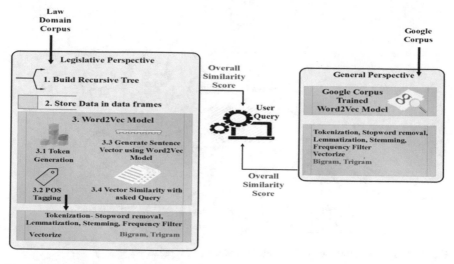

Fig. 2 Relevant subsection retrieval model

4.2 Relevant Subsection Retrieval Approach

The key objective of this research is to flourish the cognitive assistance technology area by producing fundamental advances. Relevant subsection retrieval in law domain is quite challenging problem and beneficial tool for public sector community. The approach which is discussed in this chapter is useful for both legislative community as well as client/business houses community. Stakeholders' time reduces to understand any legal matter, contracts, and regulatory. Detailed approach is discussed here.

4.2.1 Build Recursive Tree

In this initial and first process, the law domain corpus dataset is broken down into well-defined sections according to its web structure. Data crawling code has been done in such a way that sub-section comes under its parent section. This method is quite handy to deal with many recurrences. Although in data crawling sense, it is not able to cover all recurrences but that does not affect retrieval efficiency. Note that data has been scrapped and stored in recursive tree format as shown in Fig. 3 and building_recursive_tree code is mapped with the data crawling code to achieve this objective, generalized recursive tree code is given in Fig. 4.

4.2.2 Transition Tree

This step converts all the trees into python@ pandas data frame which defines the headers and their respective information separately with the naming convention of

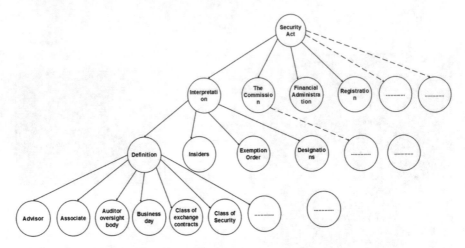

Fig. 3 Sample recursive tree of 'Security Act' content exist in law corpus

Fig. 4 Generalized
recursive tree building code

```
def building_recursive_tree(node):
    results = [node['id']]
    if len(node['children']) > 0:
        for child in node['children']:
    results.extend(building_recursive_tree(child))
        return results
```

topic headers and topic information. Following steps are taken to overcome the initial challenging task of handling and storing the unstructured data in some semi-structured format which machine can understand and comprehend.

4.2.3 Word2Vec Model

Two variations of word2vec model are applied—first Word2Vec model is used to train law domain corpus, and another is Google corpus trained Word2Vec model which uses 3.6 GB of trained corpus given by Google news data. Google corpus trained Word2Vec model is not an automated system. Further, steps are discussed which are used to train Word2Vec model for the law corpus.

- **Token generation of topic information**: Complete corpus has been converted into tokens/terms using word tokenizer of NLTK package. The base word or dictionary form of word is known as lemma, so to convert tokens to their base form, lemmatization is incorporated by removing inflectional endings and then insignificants words like stop words and tokens having length less than 2 are also removed from the law corpus content. Sample question and tokenized words are shown in Fig. 5.

Requested question:	"Does a circular have to say the debt that a director owes to the company?"
Tokenization:	['Does', 'a', 'circular', 'have', 'to', 'say', 'the', 'debt', 'that', 'a', 'director', 'owes', 'to', 'the', 'company', '?']
Tokens:	['circular', 'say', 'debt', 'director', 'owe', 'compani']

Fig. 5 Tokenized sample

While performing the tokenization the system generates tokens for topic keywords to give emphasis to the header keyword if they occur in the requested question.

For example:

Topic:	"Notice concerning notice-and-access regime recently adopted by the Canadian Securities Administrators"
Tokens:	['notice', 'concerning', 'notice', 'access', 'regime', 'recently', 'adopted', 'canadian', 'securities', 'administrators'].

- **Part-of-speech filtering**: For the remaining terms, POS tagging distribution is applied. POS graph of remaining terms of the corpus is shown in Fig. 6. We have observed that Noun—singular tag (NN), plural tag (NNS), singular proper noun (NNP), plural proper noun (NNPS) are highly significant and these tags lie maximally in corpus. Therefore, in POS filtering only useful tags are selected and terms associated with the rest other tags are removed. So, from the Fig. 5, the tokens after POS filtering are obtained as follows:

Tokens after POS filtering: ['circular', 'debt', 'director', 'owes', 'company'].

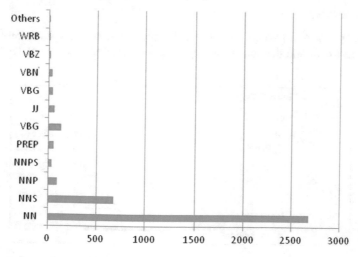

Fig. 6 Part-of-speech tag distribution graph

Word frequency filtering is applied and the words having word length less than two are removed from the corpus.

- **Generate sentence vector from Word2Vec model**: Word2Vec model is used to incorporate the semantic behavior of the corpus. Average vector of all the words in every sentence is computed. Genism@ package is used to initialize and train the model using these parameters—size, min_count, workers, window, and sample. Where, size is defined as the dimensionality of the word vectors; min_count is the minimum word count to be considered such that the total frequencies that are lower than this be ignored; workers is the number of threads to run in parallel to train the model; window is the maximum distance between the current and the predicted word in a sentence; and finally, sample is the threshold to configure down sampling of words with higher frequency. In our case, model is set for the values such as size = 300 (size of Google corpus), min_count = 10, workers = 4, window = 10 and sample = 1–3. Word2Vec represents all the terms with low dimensional vector and predict close encircling words of every term (Fig. 7).
- **Computing vector similarity**: This step calculates the vector similarity between the following two vectors—vectorized requested question, and the vectorized complete data to find the similar sub sections to the requested question that are asked by the user. This function returns a float value which represents cosine similarity and it ranges between 0 and 1. The value closer to one implies that the corresponding section is more similar.

The cosine similarity between two vectors is defined as the measure of the normalized projection of one vector over the other. It is defined by the following formula as is given in Eq. (1):

$$similarity = \cos \theta = \frac{A.B}{|A||B|} = \frac{\sum_{i=1}^{n} A_i B_i}{\sqrt{\sum_{i=1}^{n} A_i^2} \sqrt{\sum_{i=1}^{n} B_i^2}} \tag{1}$$

- **Fetching common tokens**: This step calculates the common tokens between the topic keywords and the corresponding topic information, and calculates the key-

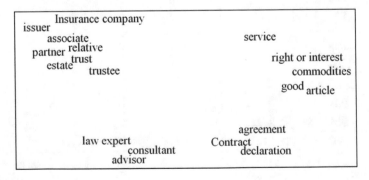

Fig. 7 Word2Vec model

word similarity. Hence, for calculating score of Word2Vec model the value of score gets calculated by adding the vector similarity score and 0.5 weightage is given to the keyword similarity. We set a threshold value of 0.4 to this score, and passed all the results above the threshold value to the TF-IDF score module.

4.2.4   Process 4: TF-IDF Score

This Module computes the TF-IDF score on Word2Vec model results. In this module, a term vector t is extracted that contains terms. Thus, for each sentence a term vector is defined as $t_i = (t_1, t_2, t_3 \dots t_x)$. While extracting terms from the corpus n-gram model is applied to find out terms having contiguous sequence of more than one term. Function is used to first identify the bigrams—a pair of consecutive words and trigrams—a group of three consecutive words.

A dictionary of bag-of-words is created to calculate the TF-IDF vector based on the occurrences of the terms in the text. It returns the vectorized form of the tokenized data. All the feature engineering, selection and pre-processing preformed in previous sections result into final feature vectors of the chosen datasets. Importance of every feature with respect to every category needs to be calculated to provide an effective recommendation. Considering the idea, measuring the importance of all features corresponding to all categories is done using a resultant dependent modified version of TF-IDF formula [9]. The modified TF-IDF is adjusted for topic wise importance of a feature. In this chapter, authors have modified TF-IDF formula according to the frequency of terms that are used in a specific topic [15, 19] as is seen in Eq. (2).

$$\frac{F}{tf} * \left(\frac{jf}{\sum jfs} \right) * \frac{log(ccs)}{cc} \tag{2}$$

here,

F frequency of term in a topic
tf total term frequency in a topic
jf total sections in a particular category
jfs total sections in the corpus
cc total topics containing a particular term
ccs total topics in the model.

Then again, similarity is calculated using the vector similarity via cosine similarity measure as is already stated in the Eq. (1).

4.2.5   Process 5: Overall Score Computation

Overall score is calculated by the summation of the two scores that are generated by the Word2Vec and TF-IDF function respectively. We give the combined threshold value of 0.55 for the combined score. Henceforth, calculate_score_word2vec function returns a pandas@ dataframe with the resultant similar topics and topic information along with the corresponding score.

5 Empirical Evaluation

As a baseline of results evaluation, we have experimented outcome at various levels which are mentioned as models in the result tables—Tables 4 and 5. Overall experiments are shown to reflect relevant subsection retrieval system performance to cover both the perspectives—user perspective and legislative stakeholder perspective. To provide a clear understanding of performance of the system, results are shown on some sample questions with respect to users and legislative stakeholder which are presented in tables—Tables 2 and 3 respectively. Both tables contain 5-5 questions each.

Few features set, and retrieval techniques combination is forming a model. Here, results are tested on five models where two features—stop word removal, n-gram and one retrieval technique—TF-IDF are applied on all the models. To determine which model contains what all features and what techniques are used in that model, all such information are presented in Tables 4 and 5 using correct (✓) and wrong (✗) sign. Table 4 shows each model result corresponding to the user questions. Table 5 shows results of the legislative stakeholder perspective.

Table 2 Sample questions asked by user

S. no.	Sample questions asked by user	Key tokens of asked question
1	What is exchange-traded security?	['exchange', 'trade', 'security']
2	What is equity investee?	['equity', 'investee']
3	How long do interim period last?	['interim', 'period']
4	Who is a SEC issuer?	['sec', 'issuer']
5	What is notice-and-access?	['notice', 'access']

Table 3 Sample questions asked by legislative stakeholder

S. no.	Sample questions asked by user	Key tokens of asked question
1	What deadline must be filed for annual financial statements?	['deadline', 'statements']
2	What will be the maximum length of the transition year?	['length', 'transition', 'year']
3	What all materials can be posted on non-SEDAR website?	['materials', 'sedar', 'website']
4	Do meeting results have to be disclosed?	['meeting', 'results']
5	When does the notice of meeting have to be issued?	['meeting']

Table 4 Goole corpus trained user questions outcome

Google corpus trained Word2Vec model parameters	Model #1	Model #2	Model #3	Model #4	Model #5
Stop-words removal	✓	✓	✓	✓	✓
Phrases: n-grams	✓	✓	✓	✓	✓
Headings with branches	✗	✓	✓	✓	✓
Level hierarchy	✗	✓	✓	✓	✓
TF-IDF (terms weightage)	✓	✓	✓	✓	✓
Synonyms	✗	✗	✓	✗	✓
POS (only nouns)	✗	✓	✓	✓	✓
Stemming and lemmatization	✗	✗	✗	✓	✓
Terms (word length > 2)	✗	✗	✗	✓	✓
Word2Vec + TF-IDF	✗	✗	✗	✗	✓
Total number of results on sample questions asked by user					
Question 1	72	35	33	20	15
Question 2	43	26	14	11	10
Question 3	13	7	2	2	1
Question 4	106	87	49	34	26
Question 5	28	13	9	7	7

Table 5 Law corpus trained legislative stakeholder questions outcome

Law corpus trained Word2Vec model parameters	Model #1	Model #2	Model #3	Model #4	Model #5
Stop-words removal	✓	✓	✓	✓	✓
Phrases: n-grams	✓	✓	✓	✓	✓
Headings with branches	✗	✓	✓	✓	✓
Level hierarchy	✗	✓	✓	✓	✓
TF-IDF (terms weightage)	✓	✓	✓	✓	✓
Synonyms	✗	✗	✓	✗	✓
POS (only nouns)	✗	✓	✓	✓	✓
Stemming and lemmatization	✗	✗	✗	✓	✓
Terms (word length > 2)	✗	✗	✗	✓	✓
Word2Vec + TF-IDF	✗	✗	✗	✗	✓
Total number of results on sample questions asked by legislative stakeholder					
Question 1	119	87	56	28	25
Question 2	36	33	28	24	23
Question 3	7	5	2	4	1
Question 4	26	19	12	8	8
Question 5	82	49	27	27	25

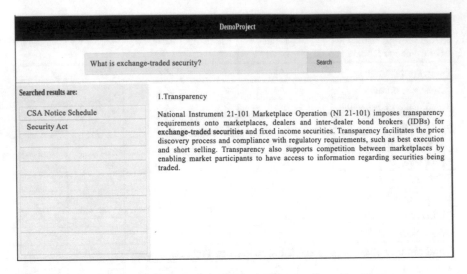

Fig. 8 Response snapshot of relevant subsection retrieval system for a user query

It is observed that as numbers of features and techniques are increasing, the outcome count corresponding to that specific query is either same or less. For example, for query 1 in Table 4, Model #1 gives 72 results, Model #2 gives 35, Model #3 gives 33, Model #4 gives 20 and finally Model #5 gives 15 outcome while containing all features and using both Word2Vec and TF-IDF techniques. Same can be observed for other user and legislative queries also.

It is clear that count of result statement is decreasing but now question is "what about performance i.e. how well the system is able to answer stakeholders' enquiry?" The result of a general/user question "what is exchange-traded security" and its corresponding response is shown in Fig. 8. Here we compute confidence score of outcome corresponding to asked law question and responses having confidence score threshold more than 80%.

6 Conclusions and Future Work

Knowledge Management (KM) system helps industries to acquire and retain business. It provides efficient and effective features to various innovative applications based on domain knowledge such as customer support, industrial innovation, insurance firms, legal firms etc. Out of various innovative applications, this chapter deals with the challenging problem of retrieving relevant and the most suitable sub-section to answer user's query by developing a sophisticated legal question answering system specifically for law domain. Legal QA retrieval system started with a focus on documents. Substantive lawyer documents are used to validate the developed system and user query searchers answer from multiple sources of information. In fact, efforts are made to provide KM support for the help of lawyers and legal clients.

Researchers' interest towards legal knowledge management is growing exponentially and surely systems that are developed using this information provide productivity to the law firm. The introduced legal knowledge management system consists of both legal documents and online available content that identify, disseminate and accumulate knowledge/information to solve legal matters. Legal document are fetched according to matching user query from online content which is available through legal knowledge sources. Further, extracted information content resources that are fetched by the system are used to build a successful legal QA system. This QA system is developed to cover two prime viewers' perspectives—Legislative stakeholder and General. Numerous NLP and IR techniques are incorporated to answer everyone's query such as Word2Vec, TF-IDF, POS tagging and N-Grams. These techniques are used to manage information resources and to provide satisfactory outcome of users' query. Experimental analysis depict that the proposed retrieval method gives promising results w.r.to precision and accuracy.

For legal domain, wide scope of research work is required where KM can provide support to complete users' various requirements. This work can further be extended in the following manner

- Dependency parsing can be used for better understanding of corpus;
- Topic modeling techniques can help to achieve better results; and
- Law domain ontology can be used to specify the law-related terminology and its interrelationships.

Acknowledgements This work is partially supported by the National Research Project CSO2017-86747-R funded by Spanish Ministry of Economy, Industry, and Competitiveness to Professor Jorge Morato, Professor of the Universidad Carlos III. Dr. Anuja Arora, Associate Professor at Jaypee Institute of Information technology is collaborated and provided technical support in this project. Law domain has many peculiarities in its syntax and vocabulary. These peculiarities have a high impact on the understanding of the texts and their recovery, which damages the rights of citizens to exercise their rights. Word embedding technology can be useful in this sense since it allows detecting those patterns that present greater difficulty for the comprehension of texts. In recent years, governments around the world are facilitating online access to perform administrative procedures, but citizens are often unable to understand the information provided. The work presented in this paper is part of the project and evaluate the incidence of these patterns with a low degree of understanding in online government documents. This paper and the mentioned project are efforts in order to alleviate this problem.

References

1. Ford VF (2012) An exploratory investigation of the relationship between turnover intentions, work exhaustion and disengagement among IT professionals in a single institution of higher education in a major metropolitan area. Doctoral Dissertation, The George Washington University

2. Ford VF, Swayze S, Burley DL (2013) An exploratory investigation of the relationship between disengagement, exhaustion and turnover intention among IT professionals employed at a university. Inf Resour Manag J (IRMJ) 26(3):55–68

3. Boella G, Di Caro L, Humphreys L, Robaldo L, Rossi P, van der Torre L (2016) Eunomos, a legal document and knowledge management system for the web to provide relevant, reliable and up-to-date information on the law. Artif Intell Law 24(3):245–283

4. Cowley J (2017) Passing a verdict: knowledge management in a top law firm. Refer 33(1):13–15

5. Curtoni P, Dini L, Di Tomaso V, Mommers L, Peters W, Quaresma P, Schweighofer E, Tiscornia D (2005) Semantic access to multilingual legal information. In: EU info workshop "Free EU information on the web: the future beyond the new EUR-LEX" of JURIX, pp 1–11

6. Heilbrun K (1997) Prediction versus management models relevant to risk assessment: the importance of legal decision-making context. Law Hum Behav 21(4):347–359

7. Wang K, Ming ZY, Hu X, Chua TS (2010) Segmentation of multi-sentence questions: towards effective question retrieval in CQA services. In: 33rd international ACM SIGIR conference on research and development in information retrieval, pp 387–394. Geneva, Switzerland. ACM

8. Gupta JP, Tayal DK, Gupta A (2011) A TENGRAM method based part-of-speech tagging of multi-category words in Hindi langauge. Expert Syst Appl 38(12):15084–15093 (Elsevier)

9. Lyytikäinen V, Tiitinen PASI, Salminen AIRI (2000) Challenges for European legal information retrieval. In: IFIP 8.5 working conference on advances in electronic government, pp 121–132

10. Radhouani S, Mottaz Jiang CL, Flaquet G (2009) Flexir: a domain-specific information retrieval system. Polibits 39:27–31

11. Pietrosanti E, Graziadio B (1999) Advanced techniques for legal document processing and retrieval. Artif Intell Law 7(4):341–361

12. Voorhees EM (2004) Overview of TREC 2004. In: TREC, pp 1–12

13. Li X, Roth D (2002) Learning question classifiers. In: 19th international conference on computational linguistics, vol 1, pp 1–7. Association for Computational Linguistics

14. Adebayo KJ, Di Caro L, Boella G, Bartolini C (2016) An approach to information retrieval and question answering in the legal domain. In: 10th international workshop on juris-information (JUTISIN), pp 1–14. Kanagawa, Japan

15. Hoque MM, Poudyal P, Goncalves T, Quaresma P (2013) Information retrieval based on extraction of domain specific significant keywords and other relevant phrases from a conceptual semantic network structure, 1–7

16. Kanapala A, Pal S (2013) ISM@ FIRE-2013 information access in the legal domain. In: 4th and 5th workshops of the forum for information retrieval evaluation, pp 1–5. ACM

17. Maxwell KT, Schafer B (2008) Concept and context in legal information retrieval. In: JURIX, pp 63–72

18. Joshi KP, Gupta A, Mittal S, Pearce C, Finin T (2016) ALDA: cognitive assistant for legal document analytics. In: AAAI fall symposium series (2016)

19. Peruginelli G (2007) Multilingual legal information access: an overview. Harmonising Legal Terminology. EURAC, Bolzano, pp 6–34

20. Quaresma P, Rodrigues IP (2005) A question answer system for legal information retrieval. In: JURIX, pp 91–100

21. Saias J, Quaresma P (2002) Semantic enrichment of a web legal information retrieval system. In: JURIX, pp 11–20

22. Turtle HR, Croft WB (1991) Inference networks for document retrieval. Doctoral Dissertation, University of Massachusetts at Amherst

23. Callan JP, Croft WB, Harding SM (1992) The INQUERY retrieval system. In: 3rd international conference on database and expert systems applications, pp 78–83. Springer, Vienna

24. Kim W, Lee Y, Kim D, Won M, Jung H (2016) Ontology-based model of law retrieval system for R&D projects. In: 18th annual international conference on electronic commerce: e-commerce in smart connected world, pp 1–6. ACM

25. Kulp S, Kontostathis A (2007) On retrieving legal files: shortening documents and weeding out garbage. In: TREC, pp 1–9

26. Moulinier I, Molina-Salgado H (2002) Thomson legal and regulatory experiments for CLEF 2002. In: Workshop of the cross-langauge evaluation forum for European languages, pp 155–163. Springer, Berlin
27. Van Opijnen M, Santos C (2017) On the concept of relevance in legal information retrieval. Artif Intell Law 25(1):65–87
28. Jain A, Arora A (2018) Named entity system for tweets in Hindi language. Int J Intell Inf Technol (IJIIT) 14(4):55–76 (IGI Global)
29. Jain A, Arora A (2018) Named entity system in hindi using hyperspace analogue to language and conditional random field. Pertanika J Sci & Technol 26(4):1801–1822
30. Jain A, Tayal DK, Arora A (2018) OntoHindi NER—an ontology based novel approach for Hindi named entity recognition. Int J Artif Intell (IJAI) 16(2):106–135
31. Walton D, Gordon TF (2005) Critical questions in computational models of legal argument. Argum Artif Intell Law 103:1–8

Printed in the United States
By Bookmasters